Behind the Wire

Kate Murphy

Behind the Wireless

A History of Early Women at the BBC

To dear Ruth
BBC women aise ♀!
love from Kate.

palgrave
macmillan

Kate Murphy
Bournemouth University
Poole, United Kingdom

ISBN 978-1-137-49171-8 (hardback)   ISBN 978-1-137-49172-5 (paperback)
ISBN 978-1-137-49173-2 (eBook)
DOI 10.1057/978-1-137-49173-2

Library of Congress Control Number: 2016936509

Printed on acid-free paper

This Palgrave Macmillan imprint is published by Springer Nature
The registered company is Macmillan Publishers Ltd. London

For Dorothy Torry (née Singer), with love

ACKNOWLEDGEMENTS

There are a great many people to thank for their part in the creation of this book. Firstly, without the Women's Development Initiative (a BBC mentoring scheme for which I was fortunate to be selected in 2000), I would not have had the bravado to request an attachment to the Diversity Unit to carry out the research for my 2002 report 'Women in the BBC: A History'. Kathy Butcher and Liz Kennedy, my mentors on the scheme, were pivotal in this. The research led to my decision to embark on a part-time PhD at Goldsmiths in 2005. My supervisor, Sally Alexander, was a total inspiration and I was also greatly helped by Vivienne Richmond. Pat Thane and Jean Seaton also offered valuable advice. I formed recipro-cal friendships with a number of women at this time, all of whom have gone on to gain doctorates or MPhils: Helen Glew, Helen McCarthy, Mari Takayanagi, Fiona Hackney, Judith Bourne, Valerie Johnson and Judy Faraday.

At the BBC, Jill Burridge, the Editor of *Woman's Hour*, was immensely supportive, essential when trying to juggle a PhD with a family and an almost full-time job. Through *Woman's Hour*, I met Juliet Nicolson who very generously allowed me to borrow the letters that Hilda Matheson wrote to her grandmother, Vita Sackville-West. Juliet is Executor of the Estate and I would like to formally acknowledge my thanks for permission to quote from the letters. More recently, I had the good fortune to meet Liz and Mark Somerville and their mother Judy Henderson, who shared with me memories of their grandmother/mother-in-law, Mary Somerville. They also loaned me a selection of illuminating private papers. Back in 2008 I was privileged to conduct an interview with Dorothy Torry (née

Singer) who joined the BBC in 1936. Dorothy, who was born in 1916, was placed in the 'Director General's office with John Reith, initially as a junior typist, and she later became personal secretary to William Haley. Her recollections of working for the Corporation added an extra dimension to my research. Dorothy (to whom the book is dedicated), supported by her daughter Clare, has continued to fill me in with snippets and details about the times.

The core of *Behind the Wireless* is the BBC archive and I would particularly like to thank Louise North at the BBC Written Archives Centre who ferreted out a vast number of important documents for me as well as continually showing enthusiasm and interest. Throughout the project I have also worked closely with Robert Seatter, the BBC's Head of History, who has always offered encouragement. At Bournemouth University, where I have worked for the past three years, I would like to thank my manager, Christa Van Raalte, for her faith and support in the project. Also my colleagues Kristin Skoog, Tony Stoller and Hugh Chignell who have offered unwavering cheer and reassurance.

I am also very grateful to Hugh Chignell and Tony Stoller for reading the manuscript of *Behind the Wireless* and offering critical advice. Angela John also took on this task, for which I am extremely thankful. I am further indebted to Angela who has always wholeheartedly supported my desire to study and write history. Others who read final versions, offering insightful comments, are Deborah Friedland, Isabel Sargent and my mother, Liz Murphy.

The extracts from BBC documents have been published with permission. I am also grateful to Oxford University Press for allowing me to draw on my article 'A Marriage Bar of Convenience? The BBC and Married Women's Work 1923–1939', *Twentieth Century British History* Vol. 25, 4, 2014, and Routledge for permission to draw on my book chapter 'From Women's Hour to Other Women's Lives: BBC Talks for Women and the Women who Made Them, 1923–1939' in Maggie Andrews and Sally McNamara, eds. (2014) *Women and the Media. Feminism and Femininity in Britain, 1900 to the Present.*

Finally, I would like to thank my long suffering family. In the time it has taken for this book to come to fruition my daughters, Louise and Emma, have long since left home and my son Corin, has grown from a small boy to an adult. My husband, David Holloway, has unfailingly been my rock, keeping me calm and on track. Thank you all.

CONTENTS

LIST OF ABBREVIATIONS

AWKS	Association of Women Clerks and Secretaries
CCSB	Central Council for School Broadcasting
DG	Director General
DIA	Director of Internal Administration
DSA	Director of Staff Administration
GO	General Office
GPO	General Post Office
LCC	London County Council
LSE	London School of Economics
NFWI	National Federation of Women's Institutes
PBX	Private Branch Exchange
WEA	Workers' Educational Association
WEF	Women's Employment Federation
WSA	Women's Staff Administrator

CHAPTER 1

Introduction

'Miss Sprott (of the BBC) takes Lunch in Hull' trumpeted the *Daily Mail* on 31 March 1939.[1] The venue for the BBC's Women's Press Representative, the local paper informed its readers, was the Women's Luncheon Club and her lecture on 'Broadcasting of Today and Tomorrow' had attracted a large audience. As an avid motorist, Elise Sprott would have invariably driven from London (avoiding main roads wherever possible) perhaps taking with her the MBE she had been awarded for services to broadcasting the previous year.[2]

Ever since I first came across the beguilingly named Elise Sprott in 2002, I have been captivated by her.[3] She was a BBC stalwart. Her first association with the British Broadcasting Company, as it was then called, was in June 1924, when she came before the microphone to give a short talk on 'Continental Fashions in Food'. The following year she was offered a £3.15s a week staff job as an Assistant in the Talks Department, where she would make her mark developing household programmes for women. Moved to the role of Women's Press Representative in 1931, she would spend the rest of her BBC career promoting the Corporation, informing women about its work (as she did in Hull) as well as telling the world about the women who worked for the BBC.

---

[1] *Hull Daily Mail*, 31 March 1939.
[2] *News Chronicle*, 29 July 1939.
[3] Kate Murphy (2002) Women in the BBC: A History 1922–2002, BBC Internal Report.

© The Editor(s) (if applicable) and The Author(s) 2016
K. Murphy, *Behind the Wireless*,
DOI 10.1057/978-1-137-49173-2_1

1

There is much about Miss Sprott that encapsulates women at the BBC in the interwar years—the subject of this book. An indomitable woman, she was in the vanguard of those whose lives had been transformed by the First World War. She started her BBC career at Savoy Hill (Head Office until the move to Broadcasting House in 1932), and then worked her way up the ranks to the senior salaried grades commanding by 1939 the generous salary of £600 a year. Without her, it is doubtful this book could have been written, or certainly much of the colour would be gone. It was her intrepidity that ensured, from late 1931, that a constant stream of features and news stories about 'The Silent Women of the BBC' appeared in the British press.[4] The hundreds of cuttings she generated, carefully catalogued at the BBC Written Archives Centre at Caversham, have provided a vital part of the jigsaw puzzle of research.[5]

Yet despite Miss Sprott's best efforts to put women in the picture, they have largely been left out of the historiography of the BBC. In fact it was a tantalising paragraph in the second volume of Asa Briggs' monumental *History of Broadcasting in the UK* that whetted my appetite for this project.[6] Here, in a few lines, Briggs touched on the 'key part' women played in the daily running of the organisation before the Second World War, including a fleeting reference to Elise Sprott.[7] Briggs was right, women worked in the interwar BBC at all levels, apart from the very top: as charwomen and kitchen hands; as secretaries and clerks; as drama producers, press officers, advertising canvassers and Children's Hour Organisers. They headed the Reference Library and the Registry; they ran the Duplicating Section and the Telephone Exchange, and, in the role of BBC Cashier, ensured wages were paid. Three women held Director-level posts: Hilda Matheson (Director of

---

[4] *Everywoman's*, February 1935.

[5] BBC WAC: P565: Personal Publicity: Press Cuttings 1924–1939.

[6] Asa Briggs (1965) *The Golden Age of Broadcasting: The History of Broadcasting in the United Kingdom*, Vol. 2 (London: Oxford University Press) pp. 457–8. The five volumes of Brigg's History include remarkably few references to women. Seaton has addressed this in her BBC history which includes a chapter on women. Jean Seaton (2015) *Pinkoes and Traitors: The BBC and the Nation 1974–1987* (London: Profile) pp. 207–31.

[7] Histories have been written of women in early broadcasting in the USA, Germany and Australia. Michele Hilmes (1997) *Radio Voices: American Broadcasting, 1922–1952* (Minneapolis: University of Minnesota Press); Kate Lacey (1994) *Feminine Frequencies: Gender, German Radio, and the Public Sphere, 1923–1945* (Michigan: University of Michigan Press); Lesley Johnson (1988) *The Unseen Voice: A Cultural Study of Early Australian Radio* (London: Routledge).

Talks, 1927–1932), Mary Somerville (Director of School Broadcasting, 1931–1947) and Isa Benzie (Foreign Director, 1933–1938) while Gweneth Freeman (always known as Miss Freeman) had responsibility for all female secretarial and clerical staff. Women were everywhere: at the staff dance, on the netball court and in the restaurant queue; their shingled hair, smart clothes and lipstick smiles a symbol of the modernity of the BBC.[8]

The BBC had been established in 1922, an auspicious time for women. The vote had been won in 1918 (or at least the partial vote, for those aged over 30) and the 1919 Sex Disqualification (Removal) Act had enabled entry to most professions. The First World War had changed the landscape in terms of women's employment and the climate of the times was one of new opportunities, particularly for the middle classes. The BBC tapped into this fresh resource. Most of the BBC's female staff were London-based (though not necessarily London-born) and those who worked at Savoy Hill/Broadcasting House are predominately the ones we shall meet in this book. After the First World War, the metropolis became the centre of Britain's economic growth and home to countless new industries, including the BBC.[9] As Sally Alexander, Selina Todd and others have shown, young women were the main beneficiaries of these industrial changes, with the middle classes now joining their working-class sisters in the labour force looking for employment that was appropriate and respectable, particularly work that was office-based.[10] The BBC with its central location, good pay and good working conditions, touched by both celebrity and grandeur, was a highly desirable place to be, at least from the mid-1920s once its status was assured. It meant, too, that the BBC could pick and choose female staff.

The BBC also attracted the unconventional. As a contributor to the staff journal *Ariel* commented, there were 'more people to the square

---

[8] As Adrian Bingham has shown, the popular press used the modernity of women as an emblem of post-war progress, both for good and bad, Adrian Bingham (2004) *Gender, Modernity, and the Popular Press in Inter-War Britain* (Oxford: Clarendon Press).

[9] See Sally Alexander (2007) 'A New Civilisation? London Surveyed 1928–1940s' *History Workshop Journal* 64, 297–316.

[10] See Selina Todd (2005) *Young Women, Work and the Family in England 1918–1950* (Oxford: Oxford University Press) pp. 1–2, 6–7; Sally Alexander (1995) *Becoming a Woman and Other Essays in 19th and 20th Century Feminist History* (New York: New York University Press) pp. 203–44. See also Teresa Davy 'Shorthand Typists in London 1900–1939' and Kay Sanderson 'Women Civil Service Clerks 1925–1939' in Leonore Davidoff and Belinda Westover eds. (1986) *Our Work, Our Lives, Our Words: Women's History and Women's Work* (London: Macmillan Education) pp. 124–44, 145–60.

inch who have had queer jobs at one time and another than there are in any other organisation'.[11] Elise Sprott was one of these. Aged 39 when she joined the BBC staff, she had been born into a Cumbrian ship-owning family and had attended private school, but not university. She then worked in motor engineering, joined a Voluntary Aid Detachment during the First World War and afterwards was appointed to the staff of Herbert Hoover's American Relief Administration, European Children's Fund.[12] It was as she began broadcasting on the BBC that she changed her name from Elsie to Elise, possibly an indication of aspiration.

In their introduction to *Women and Work Culture: Britain 1850–1950* Krista Cowman and Louise Jackson pinpoint 'social aspiration' as one of the meanings of work for women and how this could be viewed also in terms of economic necessity, self-fulfilment, vocation, duty and service, with the definition of 'skilled' or 'professionalism' adding to its status.[13] Apart from 'duty' all these motivators are apparent in the interwar BBC. Undoubtedly economic necessity was an imperative for most, if not all, female employees. Elise Sprott as a spinster, for example, had to support herself financially. But one of the most notable features of women's employment in the interwar years was its transience, a trait echoed at the BBC. This was because in the 1920s and 1930s there was a widely held assumption that once married, a woman would leave the workplace, either by custom or compulsion. Alice Head, the indomitable Managing Editor of *Good Housekeeping* magazine (which was founded the same year as the BBC) made the stark claim that it was 'perfectly easy to pick out' amongst the young women who worked in their offices, 'the ones who are filling in time until they get married, and those who are ambitious, keenly interested and anxious to make careers for themselves'.[14] At the BBC, a similar belief in 'two classes of women' would become an unspoken criterion for advancement as well as the stimulus behind a marriage bar that was introduced in 1932. However, unlike teaching, the Civil Service and banking (amongst many others professions and workplaces), where being single

---

[11] *Ariel*, June 1936.

[12] Women in Council (Journal of the National Council of Women), Summer 1962, Sprott Obituary.

[13] Krista Cowman and Louise Jackson eds. (2005) *Women and Work Culture in Britain c.1850–1950* (Aldershot: Ashgate) pp. 6–7.

[14] Alice Head (1939) *It Could Never Have Happened* (Kingswood: The Windmill Press) p. 194.

was a condition of continued employment, married 'career' women at the BBC were rarely forced to resign.[15]

As Britain's first broadcasting industry, the BBC's 'unique composition' offered the possibility of creative, administrative and technological careers, while its rapid growth meant the potential for increased responsibilities for those with the requisite drive and skills.[16] Hilda Matheson wrote of how the BBC had instigated many 'new professions', and not just for men.[17] Whether it was Kathleen Lines' innovations in the Photographic Library, Mary Candler's expanding work in radio copyright, Florence Minns' growing role as an auditioner and booker of 'talent' or, indeed, Elise Sprott's originations in *Morning Talks*, in the early years women grabbed openings for development and advancement. While the majority of BBC women were employed in the weekly paid secretarial and clerical grades, there was always the possibility of promotion to the salaried ranks where seniority brought with it improved conditions of service as well as an observable rise in prestige. This was important at a time when there were frequent warnings from the likes of Ray Strachey, Vera Brittain and Winifred Holtby, well-known (and prolific) feminist writers, about the dangers of 'dead-end' jobs.[18]

Britain was excessively class conscious in the interwar years and the BBC encapsulated the attitudes and aspirations that prevailed.[19] Because

---

[15] There are two key studies of professional women in the interwar years that relate most closely to the BBC, Alison Oram on teachers and Helen Glew on women in the GPO. Their focus is predominantly on the discrimination women faced in terms of equal pay, marital status and promotional prospects which led to impassioned political campaigning. Alison Oram (1996) *Women Teachers and Feminist Politics 1900–1939* (Manchester: Manchester University Press); Helen Glew (2009) Women's Employment in the General Post Office, 1914–1939 (Unpublished doctoral dissertation: University of London). Carol Dyhouse, in her study of women academics, found similar frustrations, echoed by Kaarin Michaelsen in her investigation of female medics. Carol Dyhouse (1995) *No Distinction of Sex? Women in British Universities, 1870–1939* (London: UCL Press); Kaarin Michaelsen (2005) 'Union Is Strength: The Medical Women's Federation and the Politics of Professionalism, 1917–30' in Cowman and Jackson, *Women and Work Culture*, pp. 161–76.

[16] *The Heterodyne*, June 1930.

[17] Hilda Matheson (1933) *Broadcasting* (London: Thornton Butterworth) pp. 45–8.

[18] Ray Strachey (1935) *Careers and Openings for Women: A Survey of Women's Employment and a Guide for Those Seeking Work* (London: Faber and Faber); Paul Berry and Alan Bishop eds. (1985) *Testament of a Generation: The Journalism of Vera Brittain and Winifred Holtby* (London: Virago).

[19] See Ross McKibbin (1998) *Classes and Cultures: England 1918–1951* (Oxford: Oxford University Press).

it employed women and men from all social classes it provides a backdrop from which to consider issues such as social mobility and appropriate spheres of work and to make gender comparisons. Working-class women and men at the BBC, for example, worked as cleaners, kitchen hands and house staff. All other waged positions required a good level of training and/or experience and there was an expectation that the BBC's office-based employees would be educated at least to School Certificate level. In consequence, it was those from aspiring working-class and lower middle-class backgrounds that predominated in weekly-paid clerical, secretarial and technical roles. The salaried grades, on the other hand, were filled by those from well-educated and wealthy backgrounds. The timbre of the BBC was overwhelmingly metropolitan and middle-class. It was also undoubtedly male-dominated, but women edged their way into many key posts and areas of work, negotiating their way around the 'public school' atmosphere and 'old boys' networks that they would have found.

One of the ways the BBC expressed its modernity was through an ethos of equality of opportunity. A non-gendered grading system operated from the start offering, in principle, equal pay and equal promotional chances. In April 1926 Reith robustly expressed his view that women Assistants should 'rank on the same footing as men'.[20] The BBC's ethos of equality appears to have existed at a handful of other forward-looking organisations in the interwar years, such as the John Lewis Partnership and the London School of Economics (LSE) and it was also apparent in professions such as advertising.[21] One of the key areas this book will address is how the newness and modernity of the BBC set it apart from traditional professions in which educated women were clustered, such as teaching and the Civil Service, where discrimination was entrenched.

This is not to say that sexual discrimination did not exist at the BBC. Practices such as the gender stereotyping of roles as well as segregation, which were the norm at this time, were evident amongst secretarial and clerical staff. For salaried women, inequalities in recruitment, discrepancies in pay and unfairness in promotion were widespread but largely hidden. This was partly because of the ad hoc way in which the Corporation

---

[20] BBC WAC:R49/940: Women Assistants 1926, Reith to All Station Directors, 30 April 1926.

[21] Judy Faraday (2009) A Kind of Superior Hobby: Women Managers in the John Lewis Partnership 1918–1950 (Unpublished MPhil dissertation: University of Wolverhampton); Ralf Dahrendorf (1995) A History of the London School of Economics and Political Science, 1895–1995 (Oxford: Oxford University Press) pp. 113–14, 235–7.

developed without set systems for recruitment, salary rises and career development, although this did slowly change towards the mid-1930s. It was also because of the individualistic nature of the BBC. Each person was treated differently and women were dispersed widely throughout the organisation, often isolated amongst men.

One of the jobs created was that of 'producer' (although it was not termed as such at this time). The BBC's *raison d'être* was to make programmes and the vast edifice of technological and administrative positions existed to support this. A small number of women were involved in the creative process itself particularly in drama, in *Children's Hour*, in music and in social documentary but, most pertinently for the female audience, in the area of women's talks. Maggie Andrews has written about the 'feminised' nature of many BBC talks as the Company/Corporation wrestled with the paradox of public broadcasting entering the private sphere of the home.[22] The 'talks aimed at women' reveal how careful consideration was given to what was considered appropriate for listeners to hear, not only in terms of the Reithian philosophy of 'inform, educate and entertain' but also with regard to the domestic environment. Should talks be about housecraft or should they take women 'out of the home'? Similar deliberations surrounded women broadcasters and their authority to speak. How far should women address women? What was considered a suitable level of expertise? In what capacity could women speak for the nation or address men?

The first five chapters of this book are about the BBC as a place for women to work. Chapter 2 provides an overview of the BBC in the 1920s and 1930s, its hierarchies and structures, the nature of men and women's work, the conditions of employment and welfare provision and Reith's relationships with female staff. The majority were waged women and Chapter 3 considers the particulars of working at the BBC in this capacity, with an eye to the pivotal role of the Women's Staff Administrator and the General Office in the recruitment, training and placement of office-based staff. Chapter 4 explores the BBC Marriage Bar introduced in 1932 and the ways in which the Corporation negotiated the application of a regressive practice in a progressive organisation. In Chapter 5 the spotlight is thrown on the BBC's salaried women and the extraordinary range of roles they occupied. Many had risen from the ranks of the weekly paid and would continue to rise into positions of authority and expertise. The BBC

---

[22] Maggie Andrews (2012) *Domesticating the Airwaves: Broadcasting, Domesticity and Femininity* (London: Continuum).

ethos of equality was most apparent here. Chapter 6 concentrates on the four highest-paid BBC women of the interwar years; the three Directors, Mary Somerville, Hilda Matheson, Isa Benzie and the Talks Assistant Mary Adams. Through a closer scrutiny of their backgrounds and their working lives the chapter unpicks the challenges and tensions that faced elite woman at this time. The final two chapters focus on broadcasting. Chapter 7 takes a chronological approach to the succession of four Talks Assistants, Ella Fitzgerald, Elise Sprott, Margery Wace and Janet Quigley, who had responsibility for the output 'aimed at women' at this time as well as highlighting the key role of Hilda Matheson in raising the profile of women's talks. Matheson, in her capacity as Talks Director, also ensured that many more female broadcasters came before the microphone and Chapter 8 considers the role of women speakers on the BBC. Their voice and their authority to broadcast were constantly contested; their appearance on the wireless largely dependent on when they spoke, what they spoke about and who they spoke to. Finally, an Epilogue gives a brief overview of women at the BBC since 1939.

1.1   RECONSTRUCTING BBC WOMEN'S LIVES

Women at the early BBC have not completely been left out of BBC histories. Hilda Matheson is widely celebrated, the subject of a biography as well as featuring in many academic accounts.[23] Mary Somerville is also acknowledged as is the work of the science producer Mary Adams and the social documentary maker, Olive Shapley. Shapley, however, is the only BBC woman from the interwar era to have written an autobiography.[24] One of the reasons why the image of the BBC is so masculine is because of the preponderance of memoirs of former employees, all men. It is these accounts, coupled with the masculine-orientated BBC documentation, which form the backbone of most research. As Jeff Hearn points out, most books and treatises on the UK's leading institutions are about men, even if they are not specified as such.[25] So while BBC documents may have been scoured for evidence about its function as an institution or as a

---

[23] Michael Carney (1999) *Stoker: The Biography of Hilda Matheson OBE, 1888–1940* (Llangynog: Michael Carney).

[24] Olive Shapley (1996) *Broadcasting: A Life* (London: Scarlet Press).

[25] This point is also made by Michael Roper and John Tosh, eds. (1991) *Manful Assertions: Masculinities in Britain since 1800* (London: Routledge) p. 3.

provider of programmes these have largely ignored women's roles. In fact, there is an extraordinarily rich seam of material that exposes the depth of women's involvement at the BBC. It is, however, widely dispersed and this book, in many ways, draws on prosopography (collective biography) to reveal the narrative of women's lives. David Hendy has argued that biography can broaden the scope of broadcasting history; that it can be enriched by an understanding of the tastes, prejudices, talents and flaws of those who worked for the BBC, grounded in their backgrounds and life experiences.[26] While for Hendy this consideration is largely in respect of programme makers, I believe it is also relevant to a broader understanding of women's place at the BBC; their motivations and aspirations and the tension between this and the realities of their daily experience of work.

Elise Sprott is a prime example of a BBC woman whose life has been reconstructed through scattered fragments. Her staff file has not been retained (as is the case with many women) instead everything we know about her is gleaned from a range of disparate sources. These might be documents held at the BBC Written Archives Centre such as memos from the Talks Department, monthly reports from the Publicity Section or the minutes of Control Board meetings. She was effusively recorded in the records of the BBC Club, is occasionally mentioned in the staff journal *Ariel* and, most evocatively, often gave interviews to newspapers and magazines as part of her job as Women's Press Representative. While these accounts need to be treated with caution (she may well have embellished her life to add interest or intrigue), they present a woman who is busy, cheery and committed and someone who clearly loved her work. A handful of allusions to the nickname 'Sprottie' add to the picture of her as an optimistic member of staff. While she is briefly referred to in Reith's expansive diaries, more telling is her almost complete absence from the letters of Hilda Matheson, with whom she worked closely for four years, an absence which hints at their fractured relationship. *Radio Times* provides topics, dates and times of her early broadcasts while the contributor files of those she brought to the wireless offer intriguing glimpses into the producer–broadcaster relationship. There is even a 25 s extract of her melodic voice recorded in 1932, gold dust from an era when hardly any sound archives of women broadcasters survive.[27] Much about Elise Sprott is still

[26] David Hendy (2012) 'Biography and the Emotions as a Missing "Narrative" in Media History: A Case Study of Lance Sieveking and the Early BBC' *Media History*, 18 (3–4), 361–78.
[27] 'The End of Savoy Hill' broadcast 14 May 1932.

unknown or contested, even her date of birth. Her obituary in the Journal of the National Council of Women (of which she had evidently become a member once she had retired from the BBC in 1945) gives this as 1883. In fact it was 1885, as her BBC staff index card confirms.

Although there are glimpses of women in Reith's diaries and the memoirs of BBC men, they are largely absent from other contemporary books. Much was written about women's employment at this time, both its problems and its opportunities, and while the BBC is occasionally cited, it is rare. Mary Agnes Hamilton was one of the many who wrote about women's work. She was also a BBC Governor (from 1932 to 1937) so it is vexing that her recollections *Remembering my Good Friends*, which includes an account of the Corporation, make only scant reference to women.[28] Hilda Matheson's *Broadcasting* (published in 1933, 18 months after her resignation from the BBC), vividly explores her views about the wireless as a positive force for democracy but again there is little about women's work within the BBC, nor does she reflect personally on her own time at Savoy Hill. For this we are fortunate that a series of intense love letters she wrote to Vita Sackville-West have survived. Written in late 1928 and 1929, when their two-year affair was at its height, these divulge the minutiae of Matheson's daily life at the BBC as she grappled with nervous politicians, recalcitrant assistants, hopeless manuscripts and absentee broadcasters. The letters, which run to around 1,000 pages, are bold in their forthright descriptions of BBC colleagues and expose the complexities of the Corporation's hierarchies and policies.

The BBC's staff newsletters and journals also burst with women. At first these were duplicated newssheets, *The Saveloy* and *The Heterodyne*, which were produced from 1928, much of the content whimsy.[29] In June 1936, the first edition of *Ariel* was published—a high-quality quarterly which included an array of articles such as Head Office and Regional news; details of individual achievements; notices of arrivals, departures and marriages; a letters page; extensive coverage of the BBC Club with its many sporting activities and, in each pre-war edition, 'Department by Department' a tour of BBC offices with vignettes of individual staff members. These, in particular, paint a vibrant portrait of life in Broadcasting

---

[28] Mary Agnes Hamilton (1944) *Remembering My Good Friends* (London: Jonathan Cape) pp. 279–88.

[29] There were two issues of *The Saveloy*, in May 1928 and Easter 1930. *The Heterodyne* was first published in May 1930 and incorporated the BBC Club Bulletin.

House in the mid-1930s: who sat next to whom, what they did, their hobbies and their indulgences. Photographs were an essential component and, beginning with the first issue, a series of 'Ariel Portraits' were commissioned which included a spectrum of BBC women from Ursula Eason, Northern Ireland's Children's Hour Organiser, to Miss Gibson, a Senior Duplicating Operator, and Mrs Starkey, the Matron. Although the settings are staged, the pictures provide an immediate impression of the individual— their age, their clothes, their style, their class. *Radio Times*, published from September 1923, occasionally included articles which touched on BBC women's work. *Radio Pictorial* (which was published from 1934 to 1939) is more significant. It was established with a specific remit to appeal to women listeners and its pages spill over with gossip and snippets about the women at the BBC.[30] Every edition had something to report, be it Isa Benzie's low-key wedding, Florence Milnes' penchant for bus travel or Mary Somerville's delight in playing with her son.

No research into early BBC history can avoid the reality that documents from the 1930s are far more prevalent than those from the 1920s. It was not until 1932 that a Written Archive was first established, reflecting awareness by the BBC that its own history was of value. Intriguingly, all the sections of the BBC connected with the retention of information were founded and headed by women. The BBC's first archivist was Kathleen Edwin while the Registry, where official documents were filed, was established in 1927 under Agnes Mills. Along with the Reference Library (under Florence Milnes), the Sound Archive (under Marie Slocombe) and the Photographic Library (under Kathleen Lines), these resources form the basis of the BBC's archival collections.

A further resource available to me was Broadcasting House itself. As a BBC employee for 24 years I would pass through the imposing brass doors of the original building almost every weekday. The foyer, for example, has changed little from the 1930s when it would have been bedecked by Mrs Webbsmith's astounding floral displays; when Caroline Towler, in full evening dress, would have greeted eminent guests before accompanying them to the studios; when charwomen would have signed in at the huge ledger which noted the time they started work. To the best of my knowledge, *Woman's Hour*'s (where I was a producer

---

[30] See Julia Taylor (2013) 'From Sound to Print in Pre-War Britain: The Cultural and Commercial Interdependence between Broadcasters and Broadcasting Magazines in the 1930s' (Unpublished doctoral dissertation: University of Bournemouth).

for 18 years) is currently located in what would have been home to the Talks Department. I also had the chance to see Savoy Hill as it would have been in the days when it was occupied by the BBC. As a guest of the IET (formerly the Institution of Electrical Engineers) I visited the run-of-the-mill red-brick block in a gutted state, as it awaited refurbishment. Using floor plans, I was able to locate, for instance, the Registry, the photographic library and the General Office and to venture along corridors where office boys would have taken Hilda Matheson's dog, Torquhil, for a walk.

This book does not claim to be a fully comprehensive account of women at the early BBC. There are many gaps. For instance, women governors and those on the Advisory Committees are only mentioned sporadically. Women who worked in the Regions are not widely covered, largely because extant documentation tends to focus on London, nor are women who came before the microphone as actresses, comediennes, singers and musicians. The latter were almost always contract staff and so had a different relationship with the BBC and is also the reason why Grace Wyndham Goldie does not feature. Her lucid critiques of radio programmes for *The Listener*, penned in the late 1930s, were contributed as a freelancer rather than as a member of staff.[31]

What this book aims to do, then, is to put the women of the early BBC centre stage. The 1920s and 30s in Britain were a period when there was a palpable buzz about new choices and opportunities, particularly for trained and educated women, and this is the first comprehensive study of women's work in a modern 'professional' industry at this time. It casts fresh light on the ways in which women 'oiled' the machine of the BBC as well as their capacity to create, to innovate and to lead. It shows that women had an important role to play in the formation of one of Britain's most influential organisations. It gives women a voice in what has, until now, largely been a history of men.

1.2   NOTES

Photographs

Because of copyright issues, it has not been possible to include photographs in the book. There are images of Hilda Matheson, Elise Sprott and Janet Quigley as part of a short article I wrote about 'Women at the BBC',

---

[31] See Charlotte Higgins for a fresh perspective on Goldie as a drama critic. Charlotte Higgins, *This New Noise*, pp. 73–5.

available on the BBC webpage: http://www.bbc.co.uk/historyofthebbc/ research/culture/women. Also in a BBC Blog I wrote on 18 September 2015.

BBC Terminology

The BBC was notorious for its use of complex job titles which were then referred to by their initials and which were constantly changed.[32] For instance the key role of Women's Staff Administrator was originally known as Women's Staff Supervisor and later as Women's Establishment Officer. For consistency, I have used Women's Staff Administrator throughout, abbreviating it to WSA.

Money and Earnings

In the interwar years, the average annual earnings for men were:[33]

Higher professional:	£582 (1924)/£634 (1936)
Lower professional:	£320 (1924)/£308 (1936)
Clerk:	£182 (1924)/£192 (1936)
Skilled manual worker:	£180 (1924)/£195 (1936)

In the interwar years, the average annual earnings for women were:

Lower professional:	£214
Clerks:	£106
Skilled manual worker:	£87

Money was in pounds (£), shillings (s) and pence (d). A pound was 20 shillings, each shilling had 12 pence. A guinea was 21 shillings.

For example, for waged staff at the BBC, annual increments were often in amounts of two shillings and six pence (written as 2s 6d or 2/6d), the equivalent, in decimal currency, of 12-and-a-half pence.

[32] Penelope Fitzgerald had fun with titles in her novel *Human Voices*, set in the BBC during the Second World War. Penelope Fitzgerald (1980) *Human Voices* (London: William Collins).

[33] Figures according to Guy Routh (1965) *Occupation and Pay in Great Britain, 1906–1960* (Cambridge: Cambridge University Press) p. 104. According to Routh earnings for 1924 and 1936 were very similar, with perhaps a slightly lower rate in 1936.

CHAPTER 2

'Growing Like a Young Giant': The BBC as a Place to Work

Lilian Taylor's arrival at the BBC in February 1923 was remarkably low-key. Reminiscing about the start of her career in an *Ariel* interview she recollected how she was shown into a small office, asked a few questions and told 'You'll do. Can you start now?' Her previous job had been for a firm of accountants who believed that 'a woman's place was the home' so the new job seemed very thrilling, 'and even more exciting it turned out!' she enthused.[1] Miss Taylor was one of the BBC's earliest recruits. When she joined as a £2.15s a week programmes clerk, broadcasting was little known in Britain outside the realm of military communication, amateur hobby or publicity stunt.[2] Even John Reith by his own admission had not known what the term meant when he was appointed to the post of General Manager of the British Broadcasting Company in December 1922.[3]

For the first ten weeks, the British Broadcasting Company operated out of two rooms loaned by the General Electric Company at Magnet House, which is where Lilian Taylor began. By all accounts this was a frenzied

[1] *Ariel*, April 1937, Lilian Taylor memories.

[2] Unlike the USA, where women were known to have dabbled in wireless as amateurs, there is little evidence of British women participating in the pre-BBC development of radio. Michele Hilmes (1997) *Radio Voices: American Broadcasting, 1922–1952* (Minneapolis: University of Minnesota Press) pp. 132–6.

[3] Reith Diaries, 14 December 1922, shows him to be 'completely mystified as to what it was all about'. For the history of broadcasting and founding of the BBC see Asa Briggs (1961) *The Birth of Broadcasting: The History of Broadcasting in the United Kingdom* Vol. 1 (London: Oxford University Press) pp. 3–142.

© The Editor(s) (if applicable) and The Author(s) 2016
K. Murphy, *Behind the Wireless*,
DOI 10.1057/978-1-137-49173-2_2

period as staff numbers swelled to more than 30 and there was great relief when more spacious accommodation was rented at the Institution of Electrical Engineers at Savoy Hill, just off the Strand. For the next nine years in these offices overlooking the Thames, the BBC would gradually expand to fill every available space; ramshackle encroachments that fitted the pioneering nature of the infant company. Then, in 1932, the opening of Broadcasting House heralded a new era for the Corporation. Around 700 staff relocated to the imposing purpose-built central headquarters of the BBC, a symbol of both authority and modernity.[4] The changing physical presence of the BBC echoed changes in the sensitivities of the workforce. At first it was a place of adventure and unpredictability, then was added glamour and prestige and finally a more sober professionalism.

During the interwar years the BBC became the pre-eminent social and cultural conduit for the nation, bringing programmes as diverse as political debates, dance band concerts and poetry readings into the home.[5] For the first time, the majority of the British people could share national events such as the FA Cup Final, the Proms and the King's Christmas Message, the immediacy of radio ensuring the scoring of goals, the lead-violin playing of Marie Wilson and the royal words of comfort were moments that could be shared.[6] Listenership grew rapidly. Within two years more than a million licences had been sold. By early 1927 this figure had doubled. Reith estimated the average number of listeners per licence was five, 'though for any special occasion an infinitely greater number can gather', so within a matter of years the audience was already many millions. At the outbreak of the Second World War, licence holders were in excess of nine million, ensuring that out of a population of 44 million, most had access to the wireless.[7] As well as generating national output from London, the BBC also operated a raft of provincial and regional stations.[8] By 1935, 85 per cent

[4] Mark Hines (2008) *The Story of Broadcasting House: Home of the BBC* (London: Merrell) pp. 38–42.

[5] Paddy Scannell and David Cardiff (1991) *A Social History of British Broadcasting, 1922–1939* (London: Basil Blackwood) pp. 277–9.

[6] Marie Wilson, a member of the BBC Symphony Orchestra, led the Proms on several occasions in the mid-1930s.

[7] Mark Pegg (1983) *Broadcasting and Society 1918–1939* (London: Croom Helm) p. 7; John Reith (1924) *Broadcast over Britain* (London: Hodder and Stoughton) p. 80.

[8] The BBC was originally run provincially, with London Station the most important of the nine initial stations. The others were Manchester, Birmingham, Glasgow, Newcastle, Cardiff, Belfast, Aberdeen and Bournemouth. The Regional Scheme, which began its slow introduction from July 1927, divided the country into seven regions: London Regional, Midlands, West, North, Wales, Northern Ireland and Scotland.

of the population had a choice of two programmes, one National and one Regional, with a twelve-hour on-air schedule each day.

As a brand new organisation, the BBC started with no set practices and its structures, its hierarchies and its policies were largely constructed on the hoof. This lack of tradition was an important element of its progressiveness. In her book *Careers for our Daughters*, Dorothy Hughes posed the question 'Where in the business world can women find special scope?' The broad general answer was 'chiefly in those businesses which are especially developments of the present century'.[9] John Reith ('Sir' John Reith from 1927, following his knighthood) played a crucial part in the establishment of these procedures. As Asa Briggs observed 'Reith did not make broadcasting, but he did make the BBC'.[10] It was Reith's vision of broadcasting and management that created and sustained the BBC throughout the interwar years. Paternalistic, dogmatic, erratic; the complexities of Reith's character have been widely unpicked, although rarely in the context of the women he worked with.[11] He undoubtedly inspired great loyalty amongst most of his staff but the centralised control he fostered, coupled with an increasingly hierarchical management system, would gradually make the BBC a less dynamic and more ponderous place in which to work.[12]

As early as 1924, Reith had made clear his belief that broadcasting should create an informed democracy, introducing men and women to an array of issues and experiences from which they had previously been excluded.[13] Hilda Matheson was also strongly of this view, her notion of 'uplift' plainly evident in her role as Director of Talks.[14] To work at the BBC was to be part of an organisation committed to bringing both enlightenment and entertainment into the home. Not only was there a sense of pride in working for the BBC but also, for many, a belief that

[9] D.W. Hughes (1936) *Careers for our Daughters* (London: A&C Black) p. 85.

[10] Briggs, *The Birth of Broadcasting*, p. 4.

[11] The main biographies of Reith are Ian McIntyre (1993) *The Expense of Glory: A Life of John Reith* (London: Harper Collins); Andrew Boyle (1972) *Only the Wind Will Listen: Reith of the BBC* (London: Hutchinson); Maritsa Leishman (2006) *My Father: Reith of the BBC* (Edinburgh: Saint Andrew Press).

[12] Asa Briggs (1965) *The Golden Age of Wireless: The History of Broadcasting in the United Kingdom*, Vol. 2 (Oxford: Oxford University Press) pp. 454–6.

[13] D.L. LeMahieu has written of how the BBC invented and projected an image of bourgeois culture and tradition that was absorbed and accepted by all levels of British society. D.L. LeMahieu (1988) *A Culture for Democracy: Mass Communications and the Cultivated Mind in Britain between the Wars* (Oxford: Oxford University Press) pp. 179–89.

[14] This is also apparent in Matheson's book about broadcasting. Hilda Matheson (1933) *Broadcasting* (London: Thornton Butterworth).

they were improving people's lives. Throughout the interwar years, the BBC sought to project a sense amongst its staff that it was more than a workplace, that they were part of a BBC 'family'. Reith believed strongly in staff welfare and paternalism, even dedicating a short section to the topic in his autobiography, *Into the Wind*.[15] At the BBC, there was a sense of camaraderie, of being special, of belonging to an organisation that was exciting and new.

Women were always a very real presence at the early BBC, making up approximately a third of the established staff. The BBC always employed more men partly because there was such a large contingency of engineers, who were similarly roughly a third of staff numbers.[16] The reasons why people chose to come to the BBC are complex, influenced by expediency, aspiration and social class. The uncommon nature of the BBC was acknowledged. Matheson described it as 'different from the routine of the civil service and equally different from the methods of a business organisation'.[17] The individualistic nature of the Corporation's employees was likewise a point of note with its 'administrative and business people' as well as 'musicians, dramatists, educationalists, novelists, journalists and artists and some who might have been dilettanti had they not found their metier' in the BBC.[18] While the term 'dilettanti' applied chiefly to men, there were also many 'characters' among the female staff. So how had this 'individualistic' organisation begun?

2.1   From Magnet House to Broadcasting House

'Pandemonium reigned!' was how the BBC official Cecil Lewis summed up the Company's beginnings at Magnet House in early 1923.[19] During these initial few weeks in tiny offices in Kingsway the BBC established the nuts and bolts of its working practices and the roots of a broadcasting schedule that included orchestral concerts, piano recitals, talks and debates, a General News Bulletin, *Children's Hour* and live opera from

[15] Reith (1949) *Into the Wind* (London: Hodder and Stoughton) p. 272.

[16] This is very apparent from BBC Staff Lists, which were introduced in 1934.

[17] Matheson, *Broadcasting*, pp. 56–7.

[18] BBC/WAC:R1/66/2: Director General Reports, Joint Memorandum by Control Board, 27 May 1930.

[19] Lewis, *Broadcasting from Within*, p. 27. Cecil Lewis was appointed at the same time as John Reith, as Assistant Director of Programmes.

Covent Garden.[20] This meant that speakers needed to be booked, musicians rehearsed, scripts typed, contracts organised, wages paid and technical hurdles overcome. It is no surprise, then, that 'the telephones never stopped ringing, the typewriters never stopped clicking, the duplicating machines duplicated for dear life', and that all would have been operated by women.[21] One of the typists was Mrs Esmond who had migrated from Marconi, a founding company of the BBC; another was Dorothy Knight who joined 'with a personal recommendation to Mr John Reith' and who confirmed that in those early pioneer days 'everybody was expected to take a hand with everything'.[22] Lilian Taylor, who opened this chapter, captured the informality of the early BBC which she described as 'growing like a young giant'. She relished the way in which everyone pitched in, for instance 'grabbing the phone' to find replacements for performers who not infrequently dropped out.[23] This hand-to-mouth existence persisted for some time, even after the move, at the end of March 1923, to Savoy Hill.

The young BBC clearly operated in an ad hoc fashion; there was no job security and the prospect of promotions and pensions would have been an act of faith. To have embarked on a venture with an uncertain future was a gamble for all those who joined in the first months. Olive May, recruited in April 1923 to set up the telephone exchange at Savoy Hill, recalled how her friends 'thought she was crazy' to leave a good job at the General Electric Company to take up the position for the BBC.[24] This spirit of enthrallment and dedication seems to have generated a strong sense of egalitarianism, a notion that everyone was in it together, working for the common good. Even the lowliest clerk and secretary believed they were doing an important job. The memoirs of BBC men who joined during this time are filled with their zeal and sense of community: the 'charming, undepartmental spirit of the place'; the 'sense of pioneering a fascinating new development'; the 'enthusiasm everyone possessed', a sentiment that was equally shared by women.[25] Richard Lambert, the future Editor of *The Listener*, believed that while this ambitious spirit lasted there was no

[20] BBC Programme Content, 1922–26.
[21] Lewis, *Broadcasting from Within*, p. 27.
[22] *Ariel* July 1958, 'Portrait of the Month: Dorothy Knight'.
[23] *Ariel*, April 1937.
[24] *Prospero*, June 1984. Prospero is the journal for BBC retired staff.
[25] Roger Eckersley (1946) *The BBC and All That* (London: Sampson Low, Marston) pp. 57–8; Peter Eckersley (1942) *The Power behind the Microphone* (London: The Scientific Book Club) p. 57; Arthur Burrows (1924) *The Story of Broadcasting* (London: Cassell) p. 72.

limit to the devotion of BBC employees 'many of whom gave their whole time and thought, in leisure as well as at the office, to the furtherance of the service'.[26]

It would not be long, however, before this sense of immersion fizzled out. The General Strike of May 1926 was possibly the last time that women and men came together for a single cause. Notwithstanding the highly contended viewpoints about the BBC's role in the strike, at Savoy Hill there appears to have been a strong sense of unity amongst staff in getting broadcasts to air.[27] At least eight women volunteered for the emergency news service for which they received high commendation; Mrs Kidson's 'flair for journalism' and Miss Milnes' 'management of routine' were particularly noted, as was the hard work of the canteen staff.[28] But by the time Maurice Gorham joined the *Radio Times* in July 1926, he was alert to an atmosphere of 'ambition, suspicion and intrigue' emblematic of an organisation undergoing rapid growth.[29] There was unquestionably a high turnover of senior male staff. The three original employees who had been recruited alongside Reith—Burrows, Lewis and Anderson—had all moved on by the mid-1920s, as had many others, particularly those who had fallen out with Reith. Mobility was far more limited for women and competitiveness was far less, and many of those who arrived at Savoy Hill would stay with the BBC for their entire career. The change from Company to Corporation in January 1927, while constitutionally significant, had a limited impact on the daily lives of staff who carried on pretty much as before.[30]

With ever burgeoning numbers, Savoy Hill as a workspace was constantly remodelled as untapped spaces, and even adjoining buildings, were requisitioned and leased.[31] It developed higgledy-piggledy into a rabbit warren of offices, rehearsal rooms, studios, libraries, laboratories and stores. A visitor to Savoy Hill might have found actors and grandees being

[26] Richard Lambert (1940) *Ariel and All his Quality* (London: Gollanz) pp. 43–4.

[27] See for example Asa Briggs (1985) *The BBC: The First Fifty Years* (Oxford: Oxford University Press) pp. 96–106; Scannell and Cardiff, *A Social History of British Broadcasting*, pp. 32–34, 108–13; Anne Perkins (2006) *A Very British Strike* (London: Macmillan).

[28] BBC/WAC:CO37: BBCo: General Strike–Staff Arrangements, Gambier Parry to Gladstone Murray, 18 May 1926; Atkinson report, 18 May 1926.

[29] Maurice Gorham (1948) *Sound and Fury* (London: Percival Marshall) p. 17.

[30] Hines, *The Story of Broadcasting House*, pp. 22–34.

[31] Floor plans in Brian Hennessy (2005) *The Emergence of Broadcasting in Britain* (Lympstone: Southerleigh) pp. 394–413.

ferried from the foyer to the Artistes Waiting Room; office boys relaying scripts, letters and memos from the typing pool to expectant production staff; receptionists dealing with opera singers, dance troupes, cabaret stars and delivery men. There was already some single-sex accommodation such as the General Office (women) and the Research Laboratory (men), while Reith ensured that executives were housed together on the second floor, his own office enjoying 'a leafy outlook' across the river.[32]

Reith's quest for a large, permanent home for the BBC had begun in early 1927 although the chosen site at Portland Place, in the little developed area of the West End to the north of Oxford Street, would take five strenuous years to come to completion. Broadcasting House when it opened in April 1932 personified the growing self-assurance and authority of the Corporation. While Magnet House and Savoy Hill were marked by amateur enthusiasm, the new building 'typified, in steel and concrete ... a new *professionalism*'.[33] As the 1932 *BBC Year Book* proclaimed, the home in which it had 'spent its childhood' had grown into 'man's estate'.[34] Eric Maschwitz, the BBC's first Variety Director, described how the 'brand-new coat-of-arms, a house-flag fluttering from its latticed mask, and a Latin inscription' overawed the humble suitors waiting in the hall.[35]

Broadcasting House was an impressive, modernist structure with 22 studios (the interior of one, the Talks Studio 3D, designed by a woman, Mrs Dorothy Trotter), a Concert Hall, dressing rooms, rehearsal rooms, lounges, libraries, echo rooms, cloakrooms and scores of offices.[36] However, it was confusing to navigate with nine floors of labyrinthine corridors arranged around a central column of studios. The starkness of the foyer was tempered by large vases of flowers, the decorative arrangements of Mrs Webbsmith (much-beloved of the popular press), which created a friendly atmosphere to help 'shy arrivals, broadcasters or visitors, feel at home'.[37] As well as new workspaces, the iconic building offered enhanced comforts for staff. It was air-conditioned and boasted an on-site Matron, Mrs Starkey, who worked from a state-of-the-art medical room. There were also two large subsidised restaurants, one which provided

[32] *BBC Year Book* 1932, p. 90.

[33] Val Gielgud (1947) *Years of the Locust* (London: Nicholson and Watson) p. 90. Gielgud's italics.

[34] *BBC Year Book* 1932, p. 98.

[35] Eric Maschwitz (1957) *No Chip on my Shoulder* (London: Herbert Jenkins) p. 70.

[36] Dorothy Warren Trotter was a qualified architect.

[37] *Radio Pictorial,* 9 October 1936.

a waitress service for up to 210 people at a time, the other a 'cold' service and the delivery of snacks.[38] The provision of medical services and cheap food were indicative of the BBC's commitment to staff welfare. The underpaid office girl lunching on a measly cup of tea and a bun was a powerful image at this time so a reduced-price meal in comfortable surroundings was a significant benefit.[39] The restaurants were also places where staff could mingle, the chance of spotting a celebrity, a dignitary or an artiste adding an extra touch of spice.

Although most of the offices were mixed, a tour of the building would have revealed, in some areas, a clear delineation between men and women's work. On the first floor much of the west side was taken up by the General Office and Duplicating Section; the scripts, staff directives, daily menus and announcer duty sheets passed from the all-female copy typists to the all-female Roneo operatives. The east side was home to the Registry where the all-female staff (apart from office boys) catalogued BBC documents, filed listener correspondence and ear-marked mail for departments and individuals. The adjoining Post Room, on the other hand, was staffed only by men. A lift ride to the seventh floor would locate the Telephone Exchange buzzing with calls, the switchboard operated by a rota of female telephonists. Above them, on the eighth floor, the L-shaped Control Room was staffed by engineers and was exclusively male. These segregated offices would have come as no surprise to a visitor. Employment in the interwar years was largely shaped by gender stereotyping and separate places of work and, in its adoption of women-only/men-only spaces, the BBC was following the convention and cultural expectations of the times.[40]

Reith, now in a bespoke oak-panelled office on the third floor overlooking Regent Street, disliked the new building and by the time staff moved in, it was already too small, necessitating many departments to be housed

---

[38] Wilfred Goatman (1938) *By-Ways of the BBC* (London: P.S. King) p. 49.

[39] See for example Mary Grieve (1964) *Millions Made my Story* (London: Gollancz) pp. 42–3.

[40] There are a large number of books which examine women's employment issues; see, for example, Sylvia Walby (1986) *Patriarchy at Work: Patriarchal and Capitalist Relations in Employment* (Cambridge: Polity Press); Jane Lewis (1984) *Women in England 1870–1950: Sexual Divisions and Social Change* (Brighton: Wheatsheaf Books); Catherine Hakim (1996) *Key Issues in Women's Work: Female Heterogeneity and the Polarisation of Women's Employment* (London: Athlone); Harriet Bradley (1989) *Men's Work, Women's Work: A Sociological History of the Sexual Division of Labour in Employment* (Cambridge: Polity Press); Albert J. Mills and Peta Tancred, eds. (1992) *Gendering Organizational Analysis* (Newbury Park: Sage).

in out-buildings.[41] For instance, within a year the adjacent St George's Hall had become home to the Variety Department while new studios opened at Maida Vale in 1934. A further key movement of staff came with the start of the fledgling television service at Alexandra Palace in North London in 1936. London was not, of course, the only home of the BBC, the provincial and later, regional staff had similar experiences of ad hoc offices being refurbished and rebuilt. Each had its own staff of managers, music executives, producers, clerks, secretaries, telephonists, caterers, engineers and so on, while a handful of the larger stations such as Cardiff, Edinburgh and Bristol engaged Women Staff Supervisors to oversee their female employees. Wherever they were based, BBC employees were part of an ever-changing and expanding hierarchy, which came increasingly under centralised control.

2.2   BBC HIERARCHIES

Amongst John Reith's first jobs as General Manager of the British Broadcasting Company was the establishment of his leadership team. One aspect was glaringly obvious; this team did not include women. His diaries tell of a constant stream of meetings with suitable and unsuitable men for positions such as station manager, chief engineer and accountant.[42] The most significant early appointee was Charles Carpendale, who joined as Reith's deputy in July 1923. Carpendale, a retired Rear-Admiral, was one of a number of military men recruited to the early BBC; his 'quarterdeck manner' and no-nonsense style reassuring to Reith.[43] Reith, who was thirty-three when he took on the BBC job, preferred senior executives who were similar to himself, mature men with military backgrounds.[44] These were the types of men he had grown up amongst, who he felt comfortable with and understood.[45] He was ambivalent at first about young university men, on the one hand appreciating their intellect and creativity,

---

[41] Reith, *Into the Wind*, p. 158.

[42] See for example, Reith Diaries, 15 February 1923, 'A succession of appointments all day ... Saw Douglas Smith, Herd, Graham and McQueen'.

[43] Gorham, *Sound and Fury*, p. 14.

[44] Hendy has written about the effect of the war on BBC staff. David Hendy (2014) 'The Great War and British broadcasting: emotional life in the creation of the BBC', *New Formations*, 82, 82–99.

[45] Annan wrote widely about this attitude. Noel Annan (1985) *Our Age: The Generation that Made Post-War Britain* (London: Harper Collins).

on the other feeling threatened by his own lack of an elite education.[46] It was not until the later 1920s that graduates predominated amongst the salaried staff.

As staff numbers grew, new structures of management were introduced. Reith believed in central control and in early 1924, with employee numbers now nearing 200, he established his Control Board, the inner-circle of his most trusted colleagues. The Control Board (which was reconstituted many times) was the executive decision-making body of the BBC. It met weekly throughout the 1920s and 30s to discuss issues that ranged from alternative wavelengths and the staffing of *Children's Hour*, to international broadcasting meetings and the use of critics on programmes.[47] One aspect remained constant; it was always men only. While it was not unknown for women to sit on the boards of management of large companies in the interwar years (for instance, in 1919, Lady Rhondda sat on 33 boards, chairing seven of them) it was comparatively rare.[48] In 1926, the BBC Control Board was formalised with five Assistant Controllers working alongside Reith and Carpendale. In 1933, with the BBC now employing more than 1,500 staff, a major structural change was introduced with the bifurcation of the Corporation into Production and Administration with two Controllers and eight heads of large departments making up the Control Board. When this proved to be unwieldy it was replaced by a new divisional structure in 1936, decision-making and policy was now in the hands of Reith, Carpendale and four divisional Controllers representing Administration, Programmes, Public Relations and Finance. This structure was maintained by Frederick Ogilvie when he took over as Director General in 1938 and was still in operation at the outbreak of the Second World War.

Although immensely powerful, Reith did not have complete autonomy in the running of the BBC. As General Manager/Managing Director of the British Broadcasting Company he was accountable to a Board of Directors

---

[46] Reith recorded, 'an interesting conversation about Etonian benefits and so on, I having said how much I regretted not having been at a real school and Varsity'. Reith Diaries, 1 February 1928.

[47] BBC/WAC:R3/3/1-14: Control Board Minutes, 12 November 1929, 5 January 1927, 9 December 1928, 2 February 1926

[48] Angela V. John (2013) *Turning the Tide: The Life of Lady Rhondda* (Cardigan: Parthian) p. 15. The journal *Time and Tide*, established by Rhondda in 1920, had an all-female board. Other women who sat on boards include Sarah Lewis at the John Lewis Partnership and Florence Sangster at the advertising agency W.S. Crawford Ltd.

who largely let him get on with the job, rubber-stamping his resolutions.[49] However, the Royal Charter which instituted the British Broadcasting Corporation in January 1927 made provision for five Governors, each to serve for five years, who expected much deeper involvement, frequently to Reith's great frustration.[50] Four out of the 16 governors who served in the interwar years were women: the socialist and political activist Ethel Snowden (1927–32); the novelist and Labour politician Mary Agnes Hamilton (1933–37); the churchwoman and Conservative activist Caroline Bridgeman (1936–39) and the penal reformer and educationalist Margery Fry (1938–39).[51] As Reith's diaries attest, he loathed Snowden and revered Hamilton and both played a leading part in issues such as the necessity for a staff association (Snowden) and the introduction of the marriage bar (Hamilton). But although the Board of Governors were important for their strategic role in approving policies, Reith and his Control Board retained overall responsibility for the day-to-day running of the BBC.

While the Control Board was significant as the meeting place for those in command, the authority of the BBC was exercised by the Controllers in their individual capacity, the agreed decisions filtering down through various divisional, departmental or section heads. Conversely, questions and issues raised by subordinates were passed up the management chain for approval. One only has to look at a BBC memo that required an answer or clarification to see how many layers of management it passed through. The larger the BBC grew the slower and more ponderous the decision-making process became.

The BBC's Administrative hierarchy was the most pertinent to women because this was the division with responsibility for all staffing issues including recruitment and pay. The Heads of Administration, including the Women's Staff Administrator (WSA), agreed annual increments, issued directives on Saturday working and implemented the marriage bar amongst a myriad of other duties. Often derided for petty regulations (for example *Ariel* reported with glee on the 'great paper clip

---

[49] Asa Briggs (1979) *Governing the BBC* (London: BBC) pp. 53–5.

[50] For Reith's relationship with his Board of Governors see Briggs, *Governing the BBC*, pp. 55–66.

[51] Reith's diaries bristle with antagonism towards Snowden. He described her as a 'truly terrible creature, ignorant, stupid and horrid', Reith Diaries, 9 March 1927. When Hamilton joined the Board, he noted that he was sure 'Mrs Hamilton would bring a livelier and more critical intelligence to bear ... So she did'. Reith, *Into the Wind*, p. 173.

war' of 1938), it was the expansion of administration that in many ways boosted opportunities for women in the BBC.[52] Increased bureaucracy, with its adherent increase in paperwork, generated more duplicating, more filing and more typing, all work for women and mostly for women in the lower grades.

Multifaceted hierarchies also existed within the ranks of women staff. As Chapter 3 will show, the WSA played a pivotal role in the lives of the BBC's waged female office workers and she had a small administrative team of her own. The female supervisors of the four women-only sections, the General Office, Registry, Duplicating Section and Telephone Exchange, in turn reported to the WSA. Fierce distinctions were also apparent within the secretarial grades with different nomenclatures, such as copy typist, shorthand typist, 'isolated' shorthand typist and secretary, denoting seniority. These hierarchies were replicated within Engineering which, apart from secretarial support staff, was an all-male division of the BBC. The convolutions of the varying types of engineers employed by the BBC (which included studio, transmitter, operations, maintenance, research, lines, equipment, installation amongst others), dictated a mind-boggling management system which was constantly in flux.[53] In 1935, an Engineering Establishment Officer was appointed specifically to oversee staffing issues, a role similar to that of the WSA. Yet, whether an engineer or a secretary, the biggest divide at the BBC was between the weekly-waged and monthly-salaried staff, a distinction that was made from the start.

2.3   BEING WAGED, BEING SALARIED

Looking back to his arrival at the BBC in the late 1920s, Maurice Gorham described the 'caste' system that he believed existed at Savoy Hill, with distinctions shown by the colour of the carpet—'blue for seniors, with mahogany furniture: the rest had grey hairline and oak'—and the way you got your tea—'juniors a cup and saucer, seniors a pot on a tray'.[54] The intricacies of the BBC's grading structures are considered at length in later chapters, particularly the repercussions in terms of pay and promotion for female employees. Even the smallest UK office in the interwar years

---

[52] Internal Instruction No. 408 forbade the use of paper clips except for mail, *Ariel*, March 1938.

[53] Edward Pawley (1972) *BBC Engineering 1922–1972* (London: BBC Books) pp. 71–8, 197–205.

[54] Maurice Gorham, *Sound and Fury*, p. 30.

would have differentiated between manager and worker but what appears unusual at the BBC is the size of the salaried workforce in comparison with the waged. This was evident even at Magnet House; out of the 31 employees who moved to Savoy Hill in March 1923 at least 12 were monthly paid. The Establishment Chart for July 1939 shows 43 per cent of the BBC's 4,435 employees were salaried, of whom 6 per cent were women. However, for comparative purposes it should be nearer 15 per cent.[55] This is because large numbers of engineers were included whom the BBC did not routinely categorise as salaried staff. If they are removed, 14.5 per cent of women were salaried at this time. This is confirmed by the Salary Information Files which contain details of all employees who held or had held monthly paid positions in the BBC prior to the Second World War and from which engineers are again omitted.[56] Out of the 830 salaried staff listed, 128 were women, 15.5 per cent of the total.

It is not known when or why the BBC introduced a grading system although this was not an uncommon practice in the interwar years. Civil Servants, for instance, were categorised into a profusion of classifications and grades.[57] At the BBC, by January 1927 all staff were categorised into eight grades: Grades A to D were salaried, while E to H comprised the waged secretarial and clerical staff. Each band denoted a different type of work, a different rate of pay and a different 'roof'. Those who performed well received an annual increment and, once at the top of the grade, there was the possibility of promotion to the next. By 1929, employee numbers had grown to such an extent that two distinct grading systems were in operation, with the letter 'W' (standing for weekly) used to differentiate between the weekly and monthly paid. The five waged grades, 'EW' (lowest) to 'AW' (highest) were further sub-divided as staffing grew. Although the weekly waged grades were often recorded separately for women and men, there was no difference in pay levels, an individual on, for example, Grade C3W was paid the same regardless of sex.[58] The waged/salaried divide did not necessarily denote social class. The BBC's waged employees

[55] BBC/WAC:R/49/178/16: Staff Policy: Establishment, 1 July 1939.

[56] BBC/WAC:R62/100/1-3: Salary Information 1923–39. The Civil Service had a combined proportion of female administrative, executive and higher clerical officers of 5.8 per cent. It did not have the anomaly of a high proportion of technicians.

[57] See, for example, Vyrnwy Biscoe (1932) *300 Careers for Women* (London: Lovat Dickson) pp. 56–66.

[58] See, for example, BBC/WAC:R49/231/1: Staff Policy: Grades and Salaries: Monthly (Except Grade 'D').

were more likely to be lower middle-class or aspiring working class but it could also be a stepping stone, especially for women graduates.[59] The BBC liked to present itself as a workplace where advancement was possible. It professed an 'office boy to Director General' mentality where an individual from the weekly paid grades could rise through the salaried ranks, often to a positon of great authority.[60]

The BBC's Director of Internal Administration, Basil Nicolls, described the Corporation's salaried staff as its 'officer' class and its ranks were evidently awash with the educated middle class.[61] The most visible demarcation was the method of being paid: wages collected by hand each week, or salaries deposited monthly into a bank account.[62] Miss Freeman, the WSA, received memos from waged women pleading to be remunerated by cheque so they could avoid the ignominy of the weekly queue at the cashier's window, but this delineation was fiercely maintained. Salaries also rose far more quickly because the standard yearly increment was much higher. For waged staff, the annual pay rise was between 2/6d and 5/-, with a few receiving perhaps 10/- or £1. For salaried staff the lowest increment was £10, rising to £100 annually for those on salaries of £1,000 or more. Being well-paid was patently advantageous.

Apart from better pay and career prospects, the BBC's salaried employees also benefitted from improved conditions of service, in particular one month's rather than one week's notice was required on either side. There were also separate toilet facilities. It can only be imagined with what joy Miss Hope Simpson opened her letter from Miss Freeman informing her that, as she was now a Controller's secretary and so promoted to the salaried grades, she could 'hang her hat and coat upstairs!', her name now added to the list of women staff permitted to use the lavatory on the

---

[59] There are scant references to the social class of BBC women, but documents show the occupation of husbands/husbands-to-be included accountant, army subaltern, school teacher, osteopath and transport foreman, all skilled jobs. BBC/WAC:R49/372:Married Women Policy: Tribunals, Freeman also stated that the BBC's clerical/secretarial women generally married 'black-coated' workers. BBC/WAC:R49/371/1:Married Women Policy:1, undated memo from Freeman c1937/38.

[60] BBC/WAC:L1/15/1: Doris Arnold staff file, Gielgud to Eckersley, 1 September 1930, comment by Carpendale.

[61] BBC/WAC:R49/31/1: Appointments Procedure:1, Report on Recruitment of Staff, 8 February 1934, Submission by Nicolls.

[62] Staff files include details of women opening bank accounts on promotion to the salaried grades.

third floor.[63] A further distinction was made for staff who earned £500 and above with perks that included First Class Rail travel, an expense account and four rather than three weeks' annual leave.

The BBC also employed a large House Staff, manual workers who, for example, cleaned studios, maintained heating and lighting systems, transported deliveries and prepared and served meals. Although the WSA was party to the welfare and conditions of service of female house staff, their recruitment and working lives were the responsibility of the House Superintendent H.L. Chilman, an ex-military man recruited to the post in 1924. House staff had different rates of pay and conditions of service from office-based staff. For those on a weekly wage, holiday leave was two weeks rather than three while hourly paid charwomen received one week's annual leave each year. Apart from promotion to supervisor level, there is little evidence of mobility, the exception being boys who were recruited to the position of 'page'. While the BBC's office-based staff might be courteous towards house staff, there is little evidence of social interaction between the two, although house staff were entitled to use the same entrances and to eat in the canteen, something 'pretty well unheard of' before the Second World War.[64]

The BBC was a workplace where women and men, although differentiated as salaried or waged, were constantly interacting with each other. Even the most senior BBC man, cocooned in his carpeted office, came into contact with many individual women each day, which was certainly not the case for some senior men in the Civil Service with their male secretaries and often exclusively male offices and departments.[65] Whether there was a culture of bullying is uncertain. Maurice Gorham averred that 'too many senior officials bullied secretaries; woe betide the liftman who had to ask one of the high-ups which floor he wanted or the new secretary who did not recognise his voice on the phone'.[66] Ralph Wade, conversely, described how a secretary could bully her chief although the power relationship here would have been very different.[67]

[63] Freeman to Miss George, Miss Hope Simpson, Miss Osborne, Miss Shawyer, 29 September 1935. BBC/WAC:L1/1699/1: Margaret Hope Simpson Staff File.

[64] The Oral History of the BBC: Interview with Mary Lewis, 2 March 1978.

[65] Alix Kilroy described herself as an object of curiosity when she joined the Board of Trade as an Assistant Principal in 1925. Alix Meynell (1984) *Public Servant, Private Women: An Autobiography* (London: Victor Gollanz) pp. 84–6.

[66] Gorham, *Sound and Fury*, p. 20.

[67] Ralph Wade (no date) Early Life at the BBC (Unpublished memoir) p. 62.

As with any organisation, there would have been good managers and bad managers but there is no indication that the BBC was worse than any other workplace. Indeed, the situation at the BBC may have been better because of its policy of recruiting trained and experienced women. This was unusual at a time when many employers preferred to take on 'girls'.

2.4   BEING YOUNGER, BEING OLDER

As the morning sun shines on the noble facade of Broadcasting House at the top of Regent Street, four to five hundred smart young women push between the big bronze doors of the building and are whizzed in high-speed lifts up to six floors of offices.[68]

This image of large numbers of young women pouring into the BBC is an apt one. The UK's female workforce in the interwar years was predominantly one of youth, predicated by women's retirement from paid work when they married.[69] Many employers preferred young female workers who were prepared to accept lower wages, on routine tasks, in jobs with few promotional prospects, particularly pertinent for the majority of British girls who finished their elementary schooling at 14.[70] At the BBC, however, there was an expectation that waged women staff would already have experience or training and the average age of joining the Corporation was 22 with few recruited before the age of 18.[71] This was very different to the Civil Service, for example, which recruited Junior Clerks, Writing Assistants and Clerical Assistants by examination, between the ages of 15 and 18.[72]

A visitor to the BBC, being escorted to a studio or a meeting, would have immediately been aware of the employment of boys. The 'boy' was

[68] *Radio Pictorial*, 27 November 1936.

[69] The 1921 Census showed women aged between 15 and 24 comprised 63 per cent of those in paid employment; in 1931, it was 69 per cent.

[70] Much has been written about young women in the interwar workforce, see particularly Selina Todd (2005) *Young Women, Work and the Family in England 1918–1950* (Oxford: Oxford University Press).

[71] These figures are derived from the 59 weekly paid women staff who left the BBC for marriage between January 1938 and April 1939. BBC/WAC:R49/371/2: Married Women Policy:2, 28 April 1939. Only two were under 18 when they started. Twelve others were aged under 20. The oldest new recruit was 29, with five aged 27.

[72] Ray Strachey (1935) *Careers and Openings for Women: A Survey of Women's Employment and a Guide for Those Seeking Work* (London: Faber and Faber) p. 2.

a time-honoured role dating back to Victorian times and at the BBC it was an established position for selected youths who had left school at 14.[73] They were supplied with milk for their health, encouraged to attend evening classes and, the BBC hoped, would qualify for junior clerical positions in the Corporation when their employment was terminated at 18.[74] Starting on a few shillings a week, office boys and pages who showed ability could—and did—rise through the BBC. W.C. Hopkins who joined as a page in 1927 (earning 15s a week), graduated to Junior Clerk, Buying Assistant and was by 1938, a Buyer on £300 a year.[75] Similarly, Wilfred Goatman who wrote a lively account of Broadcasting House in the late 1930s, started as a 12/6d 'boy' in Swansea in 1924. By 1939 he was earning £450 in Overseas Intelligence.[76] There are very occasional references to 'office girls' at the BBC. Two were recorded as working in the Duplicating Section in 1937 where they presumably would have been trained to become operatives.[77] However, the employment of young women under the age of 18 was uncommon.

While a man's age or marital status was of limited significance in terms of employment as a white-collar worker in the interwar years, for a woman it was critical. Because most women left the workforce on marriage, either by choice or enforcement, the majority of older women professionals and office workers in the 1920s and 30s, including those at the BBC, were spinsters.[78] This was compounded by the large number of 'surplus' women, estimated in the 1921 census at one and three-quarter million, who were unable to marry because of the high male casualty rate of the First World War.[79] The feminist writer Ray Strachey stressed the potential grimness of the working lives of women aged over 30 who needed to earn their own living citing, for instance, low wages, paltry savings for a

---

[73] See, for example, Samuel Cohn (1985) *The Process of Occupational Sex-Typing. The Feminisation of Clerical Labour in Great Britain* (Philadelphia: Temple University Press) pp. 205–13.

[74] BBC/WAC:R49/227: Grades and Salaries, Grades 'D' and Weekly Paid Staff, Memorandum on the Employment of Boys, 28 April 1937.

[75] Salary Information Files.

[76] Goatman, *By-Ways of the BBC*.

[77] BBC/WAC:R49/227, 19 April 1937. The girls were graded EW with a starting salary of 17/6d a week, similar to what was then paid to 'boys'.

[78] Many working-class married women continued to work; nearly all the BBC's charwomen were older married women, see Chapter 3.

[79] Virginia Nicholson (2007) *Singled Out: How Two Million Women Survived without Men after the First World War* (London: Viking) p. xi.

pension and the constant fear of being replaced by someone younger.[80] Organisations such as the Over-Thirty Association (established in 1934) and the National Spinsters' Pension Association (established in 1935) campaigned expressly for improved working conditions. The BBC, however, took a pragmatic approach towards spinsters many of whom had long careers and ultimately retired on a generous pension. Longevity of service, for instance, was rewarded with a ten-year bonus and *Ariel* in April 1937 listed 102 women, mostly spinsters, who had been with the BBC for a decade or more.[81] Many women would remain with the Corporation for the entire duration of their working lives. For example, in the summer of 1953 *Ariel* bade farewell to six women recruited in the 1920s and 30s, five of whom were unmarried.[82] Isabel Mallinson, the BBC's first Cashier, was of such value to the Corporation that she was invited to stay on past her sixtieth birthday. She was sixty-three when she retired in 1938.[83]

Concerns were sometimes expressed about the employment of older women. In 1938 the BBC's Catering Manager showed himself to be uneasy about waitresses where the 'inevitable march of time' had caused 'girls' originally selected for their general attractiveness as well as their ability, to become 'old and haggard in their appearance' as well as slower in their duties.[84] Again, a realisation that the BBC's marriage bar might create 'compulsory spinsters' caused reflection on the object of the marriage bar. Was it 'to avoid having old or oldish women on the staff?'[85] While references such as these are rare at the BBC, they confirm the derogatory way in which spinsters and older women were often viewed during this period.[86]

Plainly one woman, Gwen Williams, lied about her age. In 1937 Williams, who had been employed on a contract basis as an Accompanist for many years, was appointed to the permanent staff as a Coach.[87] On her

---

[80] Ray Strachey 'Changes in Employment' in Ray Strachey ed. (1936) *Our Freedom and its Results by Five Women* (London: Hogarth Press) p. 145.

[81] *Ariel*, April 1937.

[82] *Ariel*, Summer 1953.

[83] Salary Information Files.

[84] BBC/WAC:R49/73/1: Catering Staff, Conditions of Service, Wade to Pym, 8 September 1938.

[85] BBC/WAC:R49/372: Married Women Policy: Tribunals, Nicolls to Carpendale, 6 March 1935.

[86] Katherine Holden (2007) *The Shadow of Marriage: Singleness in England, 1914–60* (Manchester: Manchester University Press) pp. 4, 7, 40–1.

[87] BBC/WAC:L1/454: Gwendoline Williams Staff File, Dewar to Lubbock, 4 December 1936.

staff record she wrote her year of birth as 1893, in fact she was five years older, a detail only revealed in 1953 when she came to retire.[88] It is not known why Williams felt the need to disguise her true age. It is very doubtful that having offered her a job the BBC would have rescinded it if her true age had been known particularly as the salary she was offered, £450 a year, was more than that paid to her male colleagues, indicating her value to the Corporation. A more probable explanation is that she feared attitudes towards a 44-year-old woman might be less positive and it also had the effect of keeping her in work until she was 65. While a largely pragmatic approach was taken to the employment of older women whose experience and maturity might be beneficial, the predominant new recruit was an educated and trained young woman, eager to work at the BBC. She would have been well turned-out with neat hair, a hat, stockings and gloves, the clothes she wore defining her as a modern working woman.

2.5   High Heels, Low Heels

A photograph of Head Office staff taken on the steps of Savoy Hill in November 1924 shows ranks of men in suit-and-tie, some wearing spats, and a scattering of tastefully dressed women most with bobbed hair, several with long strings of beads.[89] The clothes worn to work were clearly important. They were not only a signifier of status and personal style, they also reflected the ethos of the work environment. Sally Alexander has written about the importance of dress in asserting a woman's femininity and self-esteem which would have been particularly important when she started her first office job.[90] The 'Secretary to the Editor of a Leading London Newspaper' who was called upon, in 1932, to give careers advice to girls about office work, was adamant that it seemed 'hardly necessary to say that neatness and an entire lack of anything conspicuous are the essentials of business dress'. This did not mean dowdiness either of colour or style, she insisted, 'well-cut tailored dresses, immaculate shirt jumpers and simple frocks are the right clothes to wear'.[91] Certainly in the early

---

[88] 11 December 1952. A perfunctory note indicates the embarrassment this caused.

[89] Briggs, *The Birth of Broadcasting*, p. 225.

[90] Sally Alexander (1995) *Becoming a Woman and Other Essays in 19th and 20th Century Feminist History* (New York: New York University Press) pp. 213–15.

[91] J.A.R. Cairns (1928) *Careers for Girls* (London: Hutchinson) p. 37.

years of the BBC staff wore clothes that demonstrated a strong sense of respectability. Later a touch of glamour was added, reflecting, perhaps, the Corporation's association with celebrity.[92] Decorum was always expected of both male and female staff but women, particularly weekly waged women, were subject to more petty rules and requirements.

Dress was one of the areas keenly watched over by the WSA. Clare Lawson Dick, who started as a filing clerk in 1935, recalled how 'anyone who came without stockings on a hot day would be spoken to at once'.[93] Mary Lewis, a checking clerk in Duplicating in the late 1930s remembered that 'ladies were expected to wear hats and gloves in coming to work'.[94] Dorothy Torry, who became one of Reith's junior secretaries in 1936, was appreciative of the advice she got from Miss Freeman's office about how she should do her hair and make herself look respectable.[95] There was an expectation that women would be well-turned-out, and this applied to both the weekly waged and monthly salaried staff. A photograph of a cheerful group of charwomen leaving Broadcasting House in 1934 shows them to be smartly dressed; many in decorous hats and calf-length coats, all in polished Mary-Jane shoes, at least one sporting a fur wrap.

In their manners and dress, women staff represented the dual image of the BBC—a serious organisation dedicated to public service but one that was also touched with modernity and allure. 'Millivamps!' was one visiting engineer's remark in 1931 on the 'bright young ladies' of the General Office.[96] Maurice Gorham was captivated by what he saw as the glamour of BBC secretaries, 'high-heeled, sheer-stockinged, beautifully made-up, clutching their furs around them'.[97] An alternative view is given by a *Daily Express* staff reporter in August 1937. Hiding in a doorway, he watched the BBC secretaries 'chosen for their looks as well as their efficiency', as they came out of Broadcasting House. Commenting on their appearance he noted that, 'most of them wear two-piece suits and felt Homburgs and flat shoes, all very good and very plain'.[98]

---

[92] See Carol Dyhouse (2010) *Glamour: Women, History, Feminism* (London: Zed Books) pp. 41–54.

[93] The Oral History of the BBC: Clare Lawson Dick interview, 30 March 1979.

[94] The Oral History of the BBC: Mary Lewis interview, 13 December 1978.

[95] Dorothy Torry interviewed by Kate Murphy, 28 June 2006.

[96] *BBC Year Book*, 1932, p. 90.

[97] Gorham, *Sound and Fury*, p. 23.

[98] *Daily Express*, 6 August 1937.

Gorham, who confessed to shocking his superiors by wearing flannel trousers and a red tie when he joined the Corporation in 1926, qualified his statement about dazzling secretaries with the addendum that he was surprised 'so many of them were keen on their jobs'. This viewpoint that a voguish woman could not be serious about her work is pertinent to the BBC and appears to have been adhered to by many women staff, in particular those who were salaried.[99] Photographs of the time show that those who were waged were more likely to dress glamorously—and wear make-up. As Carol Dyhouse has shown, the use of cosmetics in the inter-war years was seen as a quick and easy way to create glamour, particularly pertinent in an era that was dominated by a passion for cinema and the worship of film stars.[100] The *Daily Express* reporter, in his secret survey of BBC secretaries, noted that not one left the building with a cigarette, 'and there was only an odd one with too much make-up on her face'.[101] Although the use of cosmetics was widespread by the mid-1930s, many still saw it as unrespectable. Cecil Graves, a senior BBC executive, was caustic in his condemnation of the fair sex 'who like to dye their hair, make their nails hideous and generally obscure their natural beauty by paint and other atrocities!'[102] This viewpoint was not shared by the BBC Governor Mary Agnes Hamilton who wrote of the contemporary young woman's mouth 'exceedingly well-defined in scarlet' as the symbol of a generation who had grown up accepting female emancipation as the norm.[103]

That many BBC women wore cosmetics is apparent from a four-page spread in *Ariel* in December 1937. Headlined 'Is there a BBC Type of

[99] See Catherine Horwood (2005) *Keeping up Appearances: Fashion and Class between the Wars* (Stroud: Sutton) pp. 53–6, 67–9.

[100] Dyhouse, *Glamour*, pp. 64–5. For Alexander the use of cosmetics was a symbol of a young woman's independence and defiance at this time, Sally Alexander, *Becoming a Woman*, pp. 219–24. See also Lucy Bland (2013) *Modern Women on Trial: Sexual Transgression in the Age of the Flapper* (Manchester: Manchester University Press) pp. 108–9. Fiona Hackney has written about the growing importance of cosmetics to women's magazines in terms of articles and advertising revenue. Fiona Hackney (2011) '"They Opened Up a Whole New World": Feminism, Modernity and the Feminine Imagination in Women's Magazines, 1919–1930' (Unpublished doctoral dissertation: University of London).

[101] *Daily Express*, 6 August 1937.

[102] BBC/WAC:R51/397/1a: Talks Policy, Graves to Maconachie, 14 July 1937.

[103] Mary Agnes Hamilton, 'Changes in Social Life', in Strachey *Our Freedom*, pp. 233–5. Young working women as a symbol of modernity is also discussed in Adrian Bingham (2004) *Gender, Modernity, and the Popular Press in Inter-War Britain* (Oxford: Clarendon Press) pp. 64–5.

Feminine Beauty?' the photographs of 29 women staff reveal that most have plucked eyebrows and use lipstick; a few even have exquisitely made-up eyes.[104] Yet, the notion that to be taken seriously career-wise women should not use cosmetics is suggested by an earlier *Ariel* photographic feature where pictures of nine of the BBC's most senior secretaries (all salaried) show them to be fresh-faced.[105] Salaried women staff's distaste for cosmetics is also evident from their staff photos; those who use make-up immediately stand out. This was partly true of elite women staff, for example the only embellishment Hilda Matheson appears to have used was powder.[106] Mary Somerville and Isa Benzie, on the other hand, were definitely aficionados of lipstick, their status perhaps permitting individuality. The fact, however, that most of the BBC's salaried women chose not to wear make-up suggests that it could be viewed as frivolous and unbecoming in a woman who held a higher status job, particularly among the women themselves.

Whether to wear cosmetics was not an issue for BBC men. Here dress was more about authority and the dress-code remained largely dark suit and tie although, as the 1920s progressed, spats and pin-stripes gave way to the more up-to-date lounge suit. Photographs of senior men in *Ariel* in the late 1930s show almost all attired in formal jacket and tie. The exception was creative men, those who worked, for example, as producers and executives in Variety or Drama. Here, Gorham observed, the trend was towards 'high-necked sweaters, corduroy trousers and spongy shoes'.[107] The BBC also had a policy of appointing ex-military men as House Staff, for example as commissionaires and receptionists. They were encouraged to wear uniform and medals which would have given those entering Savoy Hill or Broadcasting House an immediate impression of propriety and discipline.

In a few areas of the BBC, a 'uniform' was expected. Announcers, as the embodiment of the BBC, were required from 1924 to wear full evening dress. For her job as night hostess, Caroline Towler wore an evening gown largely because this is what the artistes and guests she chaperoned to the Broadcasting House studios would have been wearing. Technical research staff and maintenance engineers would have worn clothes appropriate to

---

[104] *Ariel,* December 1937.
[105] *Ariel,* October 1937.
[106] Hilda Matheson Letters, 20 December 1928.
[107] Gorham, *Sound and Fury,* p. 16.

their duties, a white lab coat for the former, overalls for the latter. For the women in the Duplicating Section, a floral pinafore was provided to protect their clothes from ink while waitresses had the perk not only of a smart black dress, apron, collar and cap but also shoes, stockings and free laundry.[108]

BBC women wore clothes that reflected their self-assurance but also defined their status. For some it was a sober pullover and skirt, for others a polka-dot frock, for still others a stole or a fur. The glamour and prestige of work at the BBC, coupled with a requirement for decorum, established a mode of dress that was formal yet feminine. It also reflected the appropriateness and responsibility of the work. Seated in the front row in a 1932 photograph of the BBC Catering Department are the two women restaurant supervisors at Broadcasting House, both soberly and respectfully dressed. One of whom is Mrs Dubarry.

2.6 PERKS, CLUBS AND ASSOCIATIONS

In May 1936, Mrs Dubarry was awarded a grant of £10 towards the cost of her daughter's illness.[109] Like all staff she was eligible to apply to the BBC's Benevolent Fund for financial help in situations of ill health, bereavement and other personal matters. The Fund was characteristic of Reith's strong belief in welfarism, a strategy increasingly popular within large companies in the interwar years who used the 'extras' of clubs, pensions, subsidised catering and so on, to encourage loyalty amongst employees and to undermine potential trade union activity.[110] In 1925, the BBC had established a Provident Fund for staff which, in February 1933, was incorporated into the London Life Pension Scheme. Membership of the scheme was compulsory for all non-manual staff and optional for manual staff, the BBC matching the equivalent five per cent of salary (or optional seven and a half per cent) that was contributed. At a time when paid holidays were largely in the gift of the employer, the BBC offered a generous allowance. As early as 1925, all office-based staff who had been with the Company for more than nine months were entitled to three weeks' annual leave, house

---

[108] BBC/WAC:R49/73/1: Catering Staff, Conditions of Service, Wade to Nicolls, 11 January 1934.

[109] BBC/WAC:R3/3/11: Control Board Minutes, 26 May 1936.

[110] See, for example, Arthur J. McIvor (2001) *A History of Work in Britain, 1880–1950* (Basingstoke: Palgrave) pp. 98–9.

staff to two weeks and charwomen to a week.[111] Hours of work were also limited to eight hours a day and in 1936, half-day working on Saturdays was abandoned.[112] Long-service was also rewarded with a ten-year bonus and, for some senior employees, what was known as Grace Leave. This was three months off with a financial grant for 'creatives' who used the time to travel, to regenerate and to get fresh ideas. In addition, while the sick leave allowance was officially four weeks a year, staff with good service who were ill for longer periods were treated more kindly, with individual cases considered on merit, something which benefitted both Mary Somerville and Mary Adams.[113]

The BBC Club, introduced by Reith in 1925, was quickly embraced by staff. Ruth Cockerton, who joined as a shorthand typist in 1924 remembered Studio No.1 being used for 'Miss Osborne's physical drill class' and how swimming galas were held at the Lambeth Baths 'with Captain Peter Eckersley pirouetting on the top board'.[114] From simplistic beginnings, the Club was expanded in 1929 to embrace a purpose-built sports ground and Club House with extensive facilities at Motspur Park in Surrey, 'so complete in every respect—from plunge bath to parquet floor'.[115] The BBC Club Bulletin, first published in January 1930, listed not only a wide range of sporting activities but also an Amateur Dramatics Society and a Debating Society. By June 1930, 400 of the 700 Head Office staff were club members.[116] Elise Sprott is a prime example of a Club enthusiast. Not only was she Secretary to the Hockey Club and a keen member of the Motoring Club but she also sat on the Pavilion and Grounds Committee, where she had responsibility for catering. In December 1930, for example, 'indefatigable as always' she was praised for the Supper and Dance for the Rugby Football Club she had prepared.[117] Sprott also acted in staff plays, on one occasion, in 1936, alongside Reith in *The Sport of Kings* (she was 'Cook'; he was the butler 'Bates').[118] The BBC Club promoted 'healthy interdepartmental and individual

[111] The first Holidays with Pay Act was passed in 1938. BBC/WAC:R49/323/1: Staff Policy: Annual Leave.

[112] In summer, working hours were 9.30–5.30, in winter 9.30–6.00.

[113] The practice was commented on in an article on Welfare Arrangements at the BBC in *Nursing Times*, 28 April 1934.

[114] *Ariel,* June 1938.

[115] *BBC Club Bulletin,* January 1930. Motspur Park opened in June 1929.

[116] *The Heterodyne,* June 1930.

[117] *BBC Club Bulletin,* December 1930.

[118] *Ariel,* June 1936.

competition' and senior management were encouraged to take part.[119] In August 1930, Hilda Matheson, President of the Hockey Section, presented the winners with a Silver Challenge Cup while in November 1930 the BBC Governor Lady Snowden attended a Club Dance.[120] Separate Clubs were also established at all the BBC's Regional offices and at some Transmission stations.

Britain in the 1930s witnessed an explosion of interest in health and outdoor pursuits with organisations such as the Youth Hostel Association and the Women's League of Health and Beauty attracting large memberships.[121] The BBC, like many other large companies and organisations, tapped into this passion for sports with its profusion of sports clubs. While many were single-sex, women were ranked alongside men in motoring, rifle shooting and table-tennis and played alongside men in chess and bridge. The Midland Region even developed a Mixed Hockey team.[122] BBC teams also played outside fixtures, with the Ladies Netball Club competing against, for instance, Lloyds Bank, Southern Railway and WH Smith, while the Ladies Hockey Team took on Battersea Polytechnic and the Ministry of Labour. There was also a philanthropic side to the BBC Club which, in 1935, took the decision to fund 'BBC House', a social centre in Gateshead for the unemployed.[123]

When Broadcasting House opened in May 1932, Reith assembled the 700 Head Office staff in the Concert Hall where he reiterated that one of his main purposes was 'the health and happiness of each one of you'.[124] Reith was adamant that, because conditions of service at the BBC were considered to be good, employees did not want or need a staff association. The issue of a staff association had first been raised by the BBC governor Ethel Snowden in 1930.[125] Reith, however, with the backing of his senior managers, provided ample evidence that staff were content and did not require representation.[126] This view was not unfounded; staff were fiercely

---

[119] *The Heterodyne*, June 1930.

[120] *The Heterodyne*, August 1930, November 1930.

[121] Juliet Gardiner (2010) *The Thirties: An Intimate History* (London: Harper Press) pp. 602–3, 522–3. The BBC branch of the Women's League of Health and Beauty took part in the 1932 Hyde Park and Albert Hall demonstrations. *Heterodyne*, April–June 1932.

[122] *Ariel*, June 1936.

[123] See for example *Ariel*, June 1936.

[124] Quoted in Briggs, *The Golden Age of Wireless*, p. 465.

[125] BBC/WAC:R/1/1: Board of Governors, 4 May 1930.

[126] BBC/WAC:R1/66/2: Board of Governors, Proposed Staff Association Reports, 28 May 1930.

loyal to both Reith and to the BBC. In March 1934, for example, during a particularly ugly press campaign that claimed mass staff discontent, Lilian Taylor (now an Accounts Clerk) organised a petition expressing disgust at the allegations and asserting commitment to the BBC. It gathered 800 signatures, those of almost half the staff.[127] The issue of a staff association was discussed by the Ullswater Committee which met for many months in 1936 to consider the BBC's Charter renewal. Its final Report recommended that one should be instituted, if staff wished it.[128] A ballot held in November 1938 showed 77 per cent of staff in favour of establishing a Joint Council; however, the complexity of the negotiations meant that the first meeting did not take place until April 1941.

Although no staff association existed prior to the Second World War, BBC employees were not barred from joining unions.[129] A few areas were highly unionised; for example, many engineers were members of the Amalgamated Engineering Union while newsroom workers might belong to the National Union of Journalists, both male-dominated areas of work.[130] In general, though, unionisation within the Corporation was low and almost non-existent for women staff. Lambert believed this was because those who worked at the BBC in the 1920s and 30s were 'too middle-class to welcome trade-union notions'.[131] However, being middle-class did not preclude women from joining unions; high numbers of women teachers and civil servants were unionised at this time.[132] Rather, the lack of unionisation amongst professional women at the BBC appears to be because there were no key issues, such as enshrined unequal pay or

[127] Briggs, *The Golden Age of Wireless*, pp. 454–6.

[128] For a detailed discussion on the Ullswater Committee see Briggs, *The Golden Age of Wireless*, pp. 476–513.

[129] At the John Lewis Partnership staff were not prevented from joining trade unions but John Spedan Lewis believed that the conditions of service offered made this unnecessary. Judy Faraday (2009) 'A Kind of Superior Hobby: Women Managers in the John Lewis Partnership 1918–1950' (Unpublished MPhil dissertation: University of Wolverhampton) pp. 105–6.

[130] BBC/WAC:R49/850: Trades Unions: Amalgamated Engineering Union; R49/871: Trades Unions, National Union of Journalists.

[131] Lambert, *Ariel and All his Quality*, p. 159. Female membership of trade unions remained low in the interwar year, never reaching more than 19 per cent of total membership. Lewis, *Women in England*, p. 169.

[132] See Alison Oram (1996) *Women Teachers and Feminist Politics 1900–1939* (Manchester: Manchester University Press); Helen Glew (2009) 'Women's Employment in the General Post Office, 1914–1939' (Unpublished doctoral dissertation: University of London).

the compulsion to resign on marriage, about which senior women staff felt impassioned enough to combine.[133] Additionally, most of the BBC's waged staff were office workers, a group notoriously difficult to unionise because they mostly worked in small and disparate establishments. In 1937, the Association of Women Clerks and Secretaries launched a fierce recruitment drive at the BBC with little success, although a handful of women staff were known to be members.[134]

2.7   JOHN REITH AND BBC WOMEN

John Reith was clearly central to the structure, ethos and working conditions of the BBC. But what was his attitude towards women? Undoubtedly an imposing figure, not just in stature (he was 6ft 6ins tall, his face scarred by a war wound) but also in self-belief, his Scottish Presbyterianism was fundamental to his character as was a profound conviction, formulated at Gresham boarding school, that he would make his mark on the world.[135] Reith did not attend university, a detail that always irked him; instead he trained as an engineer, his aptitude quickly taking him to management level. His belief in a disciplined, loyal workforce, cared for by a benevolent employer, was already apparent at the Beardmore engineering works at Coatbridge in Scotland where, from 1920, he worked as General Manager.[136] Reith's experience and self-assurance evidently impressed the interview panel for the job of General Manager at the BBC for which he applied in October 1922.

The minutiae of Reith's life and BBC career can be found in his extensive diaries which offer a startling insight into his character and relationships. The most enduring association was with his mother, her photograph always had pride of place on his desk.[137] Reith had few close friends although as a young man he developed a passionate, obsessive alliance with a local boy

---

[133] In 1921, 80 per cent of female post office workers were members of the Association of Post Office Women Clerks. Glew, 'Women's Employment in the GPO', p. 192. In 1929, 83 per cent of women teachers were members of the NUT. Oram, *Women Teachers*, p. 226.

[134] BBC/WAC:R49/857: Trade Unions: Association of Women Clerks and Secretaries.

[135] In a radio interview Reith stated, 'Merely that one felt somehow or another, arrogance if you like, or some sort of conceit, that I had qualities at that age, seventeen and a half, that ought to enable me to do something in the world'. BBC Sound Archive, 87181, 'Reith Remembered', broadcast 21 June 1989.

[136] McIntyre, *Expense of Glory* p. 99.

[137] McIntyre, *Expense of Glory*, pp. 2, 24.

which would continue to haunt him throughout his years at the BBC.[138] His courtship of Muriel Odhams, whom he met in 1917 when she was a member of the Women's Legion, was awkward and unromantic.[139] They married in 1921, their son Christopher was born in 1928, their daughter Marista in 1933.[140] For Reith there was no question that his wife would give up paid work on marriage. Hers was the domestic domain, caring for their children, sorting out servants, entertaining guests, awaiting her husband's return at any time of night or day. Reith's ability to view BBC women in a different light implies a clear division in his mind between the private and the public.

There is little in Reith's background or character to suggest he would take an enlightened attitude towards BBC women staff. He had attended boarding school, served in the army and worked in engineering, all male-dominated environments. He was not intrinsically a modern man. His family background was traditional and conservative.[141] However, confronted by the enormity of broadcasting, he became aware of the need to harness the best available staff and if this included women, he was prepared to employ them. When a woman proved herself to be committed and able, he was willing to support and promote her. This forward-looking approach was encapsulated in a 1926 directive on equality of opportunity for salaried women staff.[142] Headed 'Women Assistants' it was addressed to the BBC's provincial Station Directors and required them to abandon titles such as Chief Aunt or Woman Organiser because of the limited impression of duties they portrayed. The class of women the BBC was now employing, Reith emphasised, was such that they 'should rank on the same footing as men ... [and] be as eligible as men for promotion'. Even if a large part of their activities were in connection with women and children, he stressed, this did not warrant inequality, as the efficient planning and conduct of these programmes was as important as any other. Looking to their broader responsibilities, Reith declared that there was 'no reason actually why a woman should not be a Station Director', although he acknowledged it would be 'extraordinarily difficult to find one suitable'.

---

[138] McIntyre, *Expense of Glory*, pp. 21–85.

[139] McIntyre, *Expense of Glory*, pp. 82–9.

[140] Reith's daughter, Marista Leishman's, biography of her father portrayed him as a deeply troubled man, a workaholic who dedicated his life to the BBC. Leishman, *My Father Reith*.

[141] Reith's draconian attitudes towards alcohol and divorce would become the stuff of legends.

[142] BBC/WAC:R49/940: Women Assistants, 1926, Reith to All Station Directors, 30 April 1926.

Reith's directive indicates that, at salaried level at least, he believed men and women in the BBC should be treated equally. This applied, in particular, to the highly educated women who were applying to the BBC by the mid-1920s. But this acceptance of able women did not mean he necessarily felt comfortable in their presence. He was often baffled by them. For example, he noted in his diary in April 1926 a meeting with a 'weird creature, Miss Mackenzie', who was wanted for a position in Cardiff (Margaret Mackenzie would become a senior BBC Press Officer).[143] The following year he was similarly uncertain about Miss Mills, 'a rather weird individual' who was to be the 'central filing girl' (this was Agnes Mills, a serious-minded Oxford graduate recruited to establish the Registry).[144] Reith's unease about women was captured by Lambert who offered this insight into the Director General's approach towards senior female staff:

> Sir John's attitude towards women officials in broadcasting seemed to oscillate between nervousness and sympathy. When they gained access to his presence, he found it hard to refuse them their specific requests; but afterwards, in the light of other considerations, he might find it necessary to minimise what was conceded, with a Knoxian impatience at 'the monstrous regiment of women'.[145]

Reith's hesitant attitude towards women is evident here. Yet there is little to support Lambert's view that he viewed them as a 'monstrous regiment'. Reith's frustrations and irritations with senior BBC men are well documented in his diaries; however, similar sentiments directed against BBC women are rare.[146] In his relationships with the BBC's three most senior women, Hilda Matheson, Mary Somerville and Isa Benzie, he appears to have adopted an avuncular rapport, especially in connection with Somerville and Benzie who joined the BBC from university. Reith's relationship with Matheson, although it later descended into hostility and dislike, began as one of cordiality and mutual respect.

Reith plainly took pleasure in the company of BBC women he considered to be loyal and undemanding, in particular those he helped to appoint. This included Somerville and Benzie and also, for instance, Kathleen Lines

---

[143] Reith Diaries, 29 April 1926, 28 November 1926.
[144] Reith Diaries, 22 September 1927.
[145] Lambert, *Ariel in All his Quality*, p. 69.
[146] Reith was not averse to writing damning comments about women, as his hostile references to Ethel Snowden reveal.

and Olive May. Miss May, who ran the telephone exchange, remembered Reith joining her for cocoa after hours at Savoy Hill, and he made a point of attending the Christmas celebrations of Kathleen Lines' Photographic Section.[147] Reith especially enjoyed the personal attention women gave him which included a dedicated switchboard operator and a private waitress, Mrs Swales, who prepared his cups of tea and who looked after him 'in a most maternal way'.[148] Reith was happiest either being looked after by or looking after women. He found it much harder to work with them collaboratively.

Without a doubt the closest relationships Reith developed at the BBC, whether male or female, were with his personal secretaries. He found them easy to talk to, their constant charm and dedication far more palatable and soothing than the 'stupid', 'feeble' and 'childish' behaviour he deplored in many of the men around him. Reith's biographer, Ian McIntyre, was struck by the significant role played by his secretaries, describing them as a 'succession of remarkable women who would serve, organise, advise, cosset, mother and occasionally bully Reith throughout his BBC career and beyond it'.[149] Reith built intense, fatherly, relationships with a succession of personal secretaries, Isabel Shields, Elizabeth Nash and Jo Stanley.[150] He took them to dinner and to the cinema, brought them presents from his holidays and popped in for evening coffee at their flats.[151] In return they indulged him. They entertained his mother when she was in London, telephoned him with gossip while he was away and helped choose furnishings for his office. The volume of BBC work frequently required them to stay after hours and they often toiled alongside Reith in the evenings or at weekends.[152] Reith's expectation of total loyalty meant that he was agitated when they resigned (Shields to get married in 1928, Nash to see the world in 1936), but with all of them he remained close friends. Jo Stanley accompanied Reith to Imperial Airways when he left the BBC in 1938.

[147] Reith Diaries, 23 December 1930, 21 December 1932, 20 December 1933.

[148] Interview with Dorothy Torry, 28 June 2006.

[149] McIntyre, *Expense of Glory*, p. 120.

[150] Initially, Isabel Shields worked alone, however during 1925 the complexities of Reith's job necessitated a second secretary, Elizabeth Nash, and in December 1927, a third 'assistant secretary' was recruited, Jo Stanley.

[151] See, for example, Reith Diaries, 1 February 1923, 18 April 1927, 31 August 1934, 25 April 1938.

[152] Reith Diaries, 31 January 1926. As Reith pulled together the final papers for the Crawford Committee he noted, 'Miss Shields and Miss Nash were both at the house working on evidence'.

While Reith developed close relationships with a small number of BBC women, on the whole his attitude was far more ambivalent. On the one hand he enjoyed feeling chivalrous towards them (a trait derided by Winifred Holtby as subtly undermining), on the other he found them intimidating and confusing.[153] This dual approach is discernible in other top male officials. Reith's deputy Carpendale, for example, was also uncertain about BBC women. He appreciated their value to the Company/Corporation but could be patronising, referring to even the most senior women as 'girls' (as he did the Foreign Director, Isa Benzie).[154] Men like Reith and Carpendale, older, ex-military personnel whose education and working lives prior to the BBC had been in the company of men, found it awkward to work with women as equals.[155] This was true of many in the top echelons of the Corporation. On the other hand, the young men who arrived at the BBC from the late 1920s had started their careers in a post-suffrage atmosphere and so were more accepting of working alongside women, especially their contemporaries, as further chapters will explore.

2.8   CONCLUSION

The interwar BBC was an appealing place to work. From its haphazard origins to its image as 'the prim official with a black hat and rolled umbrella' the BBC attracted bright, dedicated men and women, proud to be part of an organisation that reached into the heart—and homes—of Britain.[156] At its helm was John Reith, 'the benevolent, if strict, father' whose vision of public service broadcasting imbued the BBC with a sense of purpose and duty.[157] Although by nature a traditionalist, Reith was willing to support and promote able women who he believed were of benefit to the BBC. He also created a management framework that reflected his belief in central control. No matter how large the Corporation grew, ultimate responsibility remained in the hands of the few men at the top. As a small organisation in the mid-1920s, the decision-making process was relatively quick. But as the BBC grew and the number of management

---

[153] Winifred Holtby (1934) *Women and a Changing Civilisation* (London: Lane and Bodley Head) pp. 105–10. Noel Annan also made this point, Annan, *Our Age*.

[154] See, for example, L1/1049/2: Isa Benzie Staff File, 22 July 1937, Carpendale to?

[155] For a discussion on women as 'other' in the workplace see Rosabeth Moss Kanter (1977) *Men and Women of the Corporation* (New York: Basic Books) pp. 206–29.

[156] Gorham, *Sound and Fury*, p. 11.

[157] Freddy Grisewood (1959) *My Story of the BBC* (London: Odhams) p. 79.

layers proliferated, the Corporation became increasingly cumbersome. In many ways, however, the increased bureaucratisation of the BBC favoured the employment of women. They were the ones who typed the memos, duplicated the forms and filed the paperwork. These were waged women and here the BBC largely conformed to the customary gender stereotyping of jobs prevalent in the interwar years. Secretaries were almost invariably female and the Duplicating Section, Registry, Telephone Exchange and General Office were women-only spaces. Yet the BBC was also a place where women and men worked side-by-side and most offices were mixed.

People chose to work at the BBC for a variety of reasons. It was undoubtedly an exciting and prestigious place to be. As wireless took hold the BBC became the voice of the nation and attracted increasingly eminent people through its doors. As a member of staff, you might catch a glimpse of, or even get to work with, a film star, a sporting personality or an MP. For many, public service was the enticement, the desire to be part of an organisation that could change and influence people's lives. As one of the original employees, Cecil Lewis, pondered, the appeal of the BBC was the 'opportunity to take part in the life of the nation, to hear great men speak of their country's affairs, to become a witness of all that is said and done'.[158] The conditions of employment were also good. Women's pay was above average for the times, the work environment was congenial and perks like the BBC Club offered opportunities for socialising and recreation that generated bonhomie.

In the 1920s and 30s, women flocked to the BBC. Confidently dressed and self-assured, they included mature, experienced older women as well as the contemporary young women Mary Agnes Hamilton described as having 'the face, the brains, the general equipment' to make the best out of life.[159] It was these young women who were recruited to the BBC's waged staff as bookkeepers, clerks and shorthand typists, their experience of working for the Company/Corporation the subject of Chapter 3.

---

[158] Lewis, *Broadcasting from Within*, p. 175.
[159] Hamilton, *Our Freedom*, pp. 233–9.

'Women Who Oil the Wheels': Waged Women at the BBC

'Dear Madam, I should be very glad if you could give me an interview with a view to possible employment with the BBC', was how 21-year-old Clare Lawson Dick opened her letter to the Corporation in March 1935.[1] Brimming with optimism she went on to explain that she had always been interested in Broadcasting House and had hoped to work there one day. Her education included Channing School for Girls, the University of Grenoble, King's College London and the LSE while her previous jobs embraced both publicity work for an MP and a brief spell in charge of the Information Bureau at Harrods. She gave a phone number and informed the BBC that she would be in London the following Tuesday and Wednesday although she could travel up to town from her home in Dorking at any convenient time.

Alas, Miss Lawson Dick was to be disappointed. The frank response from the WSA, Miss Freeman, specified that unless she was qualified secretarially, there was very little prospect of obtaining a position with the Corporation, although vacancies did occasionally occur for filing clerks in the Registry.[2] Not to be deterred, Lawson Dick showered Freeman with requests for details about jobs and, after being turned down for two positions, she accepted the post of temporary

---

[1] BBC/WAC:L2/58/1: Clare Lawson Dick Staff File, Lawson Dick to Freeman, 17 March 1935.

[2] Freeman to Lawson Dick, 26 March 1935.

© The Editor(s) (if applicable) and The Author(s) 2016
K. Murphy, *Behind the Wireless*,
DOI 10.1057/978-1-137-49173-2_3

filing clerk on 10 shillings a day in May 1935.[3] A humble start for the woman who, forty years later, would become the first female Controller of BBC Radio 4.[4] Evidently well-educated (though not to graduate level), Lawson Dick was fully prepared to begin her BBC career in the weekly paid ranks.

Clare Lawson Dick's experience of getting a foothold at Broadcasting House foregrounds many issues this chapter will address: the attraction of employment at the BBC; the self-confidence of young women and their expectations of work; the requirement for secretarial training; the possibilities of promotion and the pivotal role of the WSA. Above all it highlights the two distinct categories of female staff at the BBC, the weekly and monthly paid; waged women always far outnumbering the salaried. Men were employed in waged clerical roles but most who were on a weekly wage were engineers. In 1931, the BBC employed around 200 waged female secretarial and clerical staff. By 1939 this had risen to more than 600 in jobs that encompassed shorthand and copy typists, secretaries, clerks, duplicating operators and telephonists.[5]

As the BBC grew, so did the need for increased administrative support. Not only were there more memos to type, more contracts to issue and more copyright questions to consider but there was an ever-expanding need for bookkeepers in the Accounts Office, clerks in the Gramophone Section, telephonists in the Private Branch Exchange (PBX) exchange and shorthand typists in the School Broadcasting Department. In addition, the growing number of managers created a need for more personal secretarial assistance. Larger numbers of secretarial staff necessitated an ever more complex grading system with new levels and sub-levels introduced to differentiate between the various jobs and levels of work. Starting pay varied but most began on £2–£3 a week. Those who performed well were entitled to an annual increment (usually 5 shillings) and the possibility of promotion to a higher grade. This meant potential earnings of £4 or £5

---

[3] She became Controller in April 1975. See David Hendy (2007) *Life on Air: A History of Radio Four* (Oxford: Oxford University Press) pp. 134–8.

[4] Lawson Dick's fortitude in applying for monthly paid posts paid off in 1942. After six failed interviews and two sideways moves, she was promoted to a Grade D salaried Assistant in Programme Planning.

[5] The BBC Staff List 1934 shows 210 shorthand typists, 35 secretaries, 24 copying typists, 71 female clerks, 24 telephonists, 15 duplicating operators, 1 female book-keeper, 1 female cheque-writer and 1 office girl.

a week were within the reach of many BBC women at a time when £3 a week was seen as a good wage.[6]

Young women took up office work in increasingly large numbers in the interwar years.[7] London in particular was a magnet to those eager for independence and excitement; sharing flats and bedsitters and seeking out decent employment that might offer opportunities to save money or finance a carefree lifestyle. For many, as Chapter 4 will show, marriage was the objective so there was little thought of a long-term career. Clare Lawson Dick wrote an article about this in *The Lady* in 1938 (prior to the BBC she had also attended journalism school).[8] Entitled 'Leaving Home' she began with a statement of how 'thirty years ago it was a shocking thing for a girl to leave home and earn her own living'. Lawson Dick then asserted that 'Today parents are rightly proud of a daughter who wants to fill in the time before marriage usefully'. That this was aimed at the middle-class reader is plain; working-class girls would always have worked. But it hints at the new freedoms that were available to young women. Work, for many, was about leaving the family home, doing something that comfortably passed the day, being with friends and perhaps even finding a husband.

Social advancement might also be the goal. As Judy Giles points out in the *Parlour and the Suburb*, young women were eager to distance themselves from the lifestyle of their mothers, looking to modern-style living and companionate marriages.[9] In her study of female Civil Service clerks in the interwar years Teresa Davey revealed how bright girls from working-class families, by passing the competitive entry exam, were able to achieve this social mobility to the lower middle classes. The job not only rewarded them with higher status but enabled

---

[6] Ray Strachey (1935) *Careers and Openings for Women: A Survey of Women's Employment and a Guide for Those Seeking Work* (London: Faber and Faber) p. 117.

[7] See, for example, Gregory Anderson, ed. (1988) *The White Blouse Revolution: Female Office Workers since 1870* (Manchester: Manchester University Press) pp. 11–14.

[8] *The Lady*, 19 May 1938. Lawson Dick attended the London University Journalism School. See Fred Hunter (2012) *Hacks and Dons: Teaching at the London Journalism School 1919–1939: Its Origin, Development and Influence* (Colchester: Kultura Press) pp. 113, 212. She was given permission by the BBC to write the article.

[9] Judy Giles (2004) *The Parlour and the Suburb: Domestic Identities, Class, Femininity and Modernity* (Oxford: Berg) pp. 47–64.

them to 'marry-up'.[10] The BBC, with its middle-class ethos, similarly offered the chance of social advancement to those fortunate enough to get through its doors. The feminist writer Ray Strachey, while warning of the 'unsatisfactory and dead-end' nature of much office work, did concede that for those who were trained and talented, there were possibilities for high wages and an interesting, challenging occupation, something evidently available at the BBC.[11]

At the BBC, while the actual job of the copy typist or duplicating operator might be monotonous, the work environment was far from dull. Talks scripts, school broadcasting pamphlets and publicity material, for instance, had an intrinsic interest, and much was confidential. Page two of the 'Instructions to Women Clerical Staff', for instance, included a clause about leakage of information and the dangers of passing on office gossip.[12] Production secretaries and telephonists would have been in direct contact with actors and prominent public figures; for others there might be the chance to pass a sports star or famous musician in reception areas or in lifts. Waitresses too, could be touched by celebrity although the status of catering staff, as was the case for all House Staff, was different to that of the typist or clerk.

The BBC was one of the workplaces that Strachey would have described as an 'exceptional' place for waged women to work. By this she meant offices where there were prospects for advancement, where pay could rise above £3 a week and where women beyond marital age were not discriminated against.[13] As *Radio Pictorial* succinctly put it, 'possibly the glamour … attracts the women, or maybe it is just that the BBC is known to be a good employer'.[14]And because of its desirability the BBC could be selective about its female staff.

---

[10] Teresa Davy 'Shorthand Typists in London 1900–1939' in Leonore Davidoff and Belinda Westover, eds. (1986) *Our Work, Our Lives, Our Words: Women's History and Women's Work* (London: Macmillan Education) pp. 145–59.

[11] Strachey, *Careers and Openings*, pp. 137–8. This viewpoint was also shared by Vera Brittain (1928) *Women's Work in Modern England* (London: Noel Douglas) pp. 51–2. Nicholson has also written about the reassurance of a good clerical job, Virginia Nicholson (2007) *Singled Out: How Two Million Women Survived without Men after the First World* War (London: Viking) pp. 112–13.

[12] 'Instructions to Women Clerical Staff', issued by Freeman, 6 July 1937.

[13] Strachey, *Careers and Openings*, p. 118.

[14] *Radio Pictorial*, 23 April 1937.

3.1   GETTING AN OFFICE JOB AT THE BBC

In April 1938, *Radio Pictorial* ran a two-page spread, purportedly written by a £3.10s-a-week typist, in which she described how she came to
work at the BBC.[15] Before she joined the Corporation, the unnamed typist
explained, she had held 'two decent commercial jobs', as an office junior
and as a secretary in a shipping office. When a slump in the shipping world
made her job precarious, she looked for another, seeing the BBC position
advertised in the 'smalls'. Aware of the need to look smart for her interview, she chose to spend £4 of her savings on a 'neat grey costume and a
pair of 8s 11d stockings—the most I'd ever paid for stockings till then'. It
was a wise decision as the impression she made was good. Questions asked
by the staff committee included details of her school, her examinations,
her shorthand speed and her family life, although they did not ask for
references. 'After all', she surmised, 'Some 2,000 secretaries have already
been chosen and they get to know how to sum up a girl'.

It is probable that the 'unnamed typist' was recruited under an experimental scheme introduced by the Corporation in 1937 whereby advertisements were placed twice-yearly in the London and main provincial papers.[16]
Prior to this the BBC had no fixed method of appointing secretarial and
clerical staff. Unlike the Civil Service, with its age-specific entry examinations for female clerks and typists, the BBC used a variety of ad hoc ways to
fill vacancies. The earliest recruits were almost certainly engaged through
word of mouth, a method gradually augmented by casual applications and
by recommendations to the WSA from secretarial training establishments
and employment bureaus.

The 'unnamed typist' had held two commercial positions prior to her BBC
appointment and the BBC undoubtedly had a preference for experienced
secretarial staff. In 1936, for example, according to *Ariel* Mary Campbell, a
filing clerk with responsibility for the Registry in Edinburgh, had spent three
years at the National Library of Scotland prior to her arrival at the BBC while
Nancy Lyons, a clerk in the Press Department had previously worked in a

---

[15] *Radio Pictorial*, 8 April 1938. The article explained that it was against BBC rules for
employees to write about their jobs without permission, 'but we have received this interesting account, written by a responsible radio journalist, as told to him in various interviews and
conversations, which we publish for undoubted interest'.

[16] BBC/WAC:R49/561/1: Recruitment of Women Clerical Staff, 10 November 1936.
Selected applicants were interviewed by an Appointments Board that included the WSA or
her Deputy.

solicitor's office.[17] Similarly, three secretaries in the Programme Department in Birmingham, Dorothy Ryland, Elsie Lewis and Margaret Hock (who spoke four languages), had respectively taught elocution and games to a class of 48 children; held the fort for a stockbroker, and carried out secretarial work at Studley Agricultural College.[18] As the 1938 *BBC Handbook* stated in a paragraph on Staff Recruitment, 'girls over twenty, with good experience in a previous job, are preferred to girls of eighteen or nineteen straight from their training college with no practical experience'.

While the BBC preferred experienced staff, all secretarial recruits were required to have a good general education coupled with a high standard of shorthand and typing.[19] This would have militated against working-class girls who left elementary school at 14, although it was possible, and encouraged, to improve oneself through attending evening classes. Because the BBC did not offer training (it was not until the Second World War that the BBC set up its own Secretarial Training School) it relied heavily on secretarial colleges which proliferated in the interwar years. In her careers advice book, Ray Strachey was adamant that parents should be very careful when selecting a training school for their daughter. Reputable establishments were costly and would only have been available to girls from well-to-do homes.[20] To give an example, an advertisement for Miss Kerr Sanders Secretarial Training College was aimed at 'well-educated girls wishing to qualify for the higher branches of the secretarial profession'.[21] The BBC was in touch with many of the top concerns in London included Kensington College, St James Secretarial College and the Triangle Secretarial College along with agencies run by well-regarded individuals such as Mrs Hoster, Miss Kerr Sanders and Dilys Ajax.[22]

Dorothy Torry (née Singer) and Hilary Cope Morgan joined the BBC in a secretarial capacity at this time.[23] Both were middle-class young

---

[17] *Ariel*, June 1936.

[18] *Ariel*, December 1936.

[19] This was 120 words-per-minute shorthand and 60 words-per-minute typing, for which they were given a test.

[20] Strachey, *Careers and Openings*, p. 141.

[21] Central Employment Bureau for Women and Careers Association (1931) *Careers and Professional Training: A Guide to Professions and Occupations for Educated Women and Girls* (London: The Women's Employment Publishing Company) p. xvi.

[22] BBC/WAC:R49/31/1: Appointments Procedure: Report on Recruitment of Staff, Ernest Barker to John Reith, 8 February 1934.

[23] Interview with Dorothy Torry conducted by Kate Murphy, 28 June 2006; interview with Hilary Cope Morgan, conducted by Elizabeth McDowell, 11 November 2006.

women who had left school at seventeen and attended secretarial training school in London: Dorothy Singer at Kensington College (chosen by her parents because it was residential) and Hilary Cope Morgan at Miss Kerr Sanders (while living with her grandmother). Miss Singer, after finishing her course, worked briefly at the College and then as a secretary before, aged nineteen, she wrote to the BBC in 1936 about possible openings. She was invited for an interview and was offered a job as a typist on £2.10s a week. Miss Cope Morgan's BBC interview was arranged by Miss Kerr Sanders herself, who decreed she should wear a hat and gloves. Asked about her interests and background and whether she had played hockey or lacrosse at school, she was recruited, at the age of 20, as a typist in May 1939.

The BBC expected hard work and commitment from female secretarial and clerical staff. High standards of decorum were demanded, for instance first names were rarely used and strict punctuality in the morning and on the return from lunch were observed.[24] Clare Lawson Dick recalled how one girl was reprimanded for walking through the door of a popular coffee shop near the BBC, without standing back to allow a more senior secretary to go through first.[25] This emphasis on propriety extended to other personal matters such as the attitude towards boyfriends, one newspaper commenting that they were not allowed to wait in the building but lurked 'in the shadows of the doorways of the church over the way and of Queen's Hall'.[26] How women staff felt about these rules is hard to gauge because there was no official forum for discontent, but the gentle mocking of inconsequential regulations was routine in *Ariel*. Behind the directives and with responsibility for the centralised control of the BBC's female office staff, was the WSA.

3.2   THE WOMEN'S STAFF ADMINISTRATOR

'Meet, as the Americans say, Miss Freeman' was the invitation from *Woman's Own* to its readers; 'She Rules the Roost in Broadcasting House', proclaimed the *Daily Express* while the *Daily Mirror* asserted, 'She's behind the girls behind the programmes!'[27] As these headlines attest,

[24] 'Instructions to Women Clerical Staff'.
[25] The Oral History of the BBC: Clare Lawson Dick interview, 30 March 1979.
[26] *Radio Pictorial*, 8 April 1938.
[27] *Woman's Own*, 21 January 1933; *Daily Express*, 31 August 1935; *Daily Mirror*, 11 January 1936.

Freeman, the WSA, was critical to women's work at the BBC. According to the *Morning Post*, it was her 'shrewd judgement in selecting women staff' that had helped towards 'women's success in wireless'.[28] The position had existed since February 1923 when Caroline Banks was appointed Women's Staff Supervisor to oversee the General Office at Magnet House. Banks held the position until her marriage in 1931 when she was replaced by Freeman who retained the post until 1942.

The employment of a senior woman to oversee female staff, especially in terms of welfare, was not uncommon in interwar Britain. Lever Brothers, Boots and Sainsbury's, for instance, all had women staff supervisors.[29] At the Bank of England, the Superintendent of Women Clerks dealt with 'discipline and all feminine matters including dress, medical affairs, report cards, resignations and marriage gratuities'. Olga Collett, who held this position at ICI, described in a radio talk her responsibility for the care and control of 650 women as 'a blend of mother, schoolmistress, employers' representative and Hand of Fate'. She acknowledged that she had the 'rather terrifying power to influence and decide the course of many young lives', a description that could equally have been applied to Freeman.[30] The position of woman supervisor was the legacy of the feminisation of clerical work. When women first entered offices as clerks and secretaries in the late nineteenth century it was seen as prudent for them to be accommodated separately from men. Not only were they physically segregated but they were usually employed on specific tasks considered suitable for their sex.[31] In consequence, organisations such as banks and the Civil Service developed the role of female supervisor to oversee women-only offices and areas of work.[32] At the BBC the position would develop into one of immense authority although it was never acknowledged with the title of department head.

Why Reith opted to recruit a Women's Staff Supervisor in 1923 is unknown, but the importance attached to the position is observable in

---

[28] *Morning Post*, 18 May 1936.

[29] Elizabeth Hennessy (1992) *A Domestic History of the Bank of England, 1930–1960* (Cambridge: Cambridge University Press) p. 328.

[30] 'Other Women's Lives', broadcast 29 May 1937.

[31] See, for example, Anderson, *The White Blouse Revolution*, pp. 36, 58–63.

[32] The position also existed in industry. See, for example, Miriam Glucksmann (1990) *Women Assemble: Women Workers and the New Industries in Inter-War Britain* (London: Routledge) p. 108; Clare Wightman (1999) *More than Munitions: Women, Work and the Engineering Industries 1900–1950* (London: Longman) p. 155.

Caroline Banks' interview; she was quizzed by three senior managers including Reith.[33] Little is known of Banks' background apart from her age, 26, and that she lived with other career women in Russell Square.[34] Neither is it known whether she had supervised female staff before. What is certain is that she was initially employed to oversee a dozen female staff. When she left on marriage in 1931, she was responsible for around 200 women. Her income had also risen considerably. From a weekly wage of £3.5s, she was promoted to the salaried staff in April 1925 on £260 a year. By her resignation this had risen to £425 per annum.[35] Reith wrote a heartfelt valedictory applauding her role in the selection, training and promotion of women clerical staff and her responsibility for their working conditions and efficiency.[36] Through her work, Reith emphasised, she had made 'a material contribution ... to the establishment of the BBC'.

Bank's duties were never explicitly written down so her exact role is unclear. However, it included interviewing and appointing women staff, making decisions about their placements, overseeing their annual reports, influencing their wage rises, sanctioning their wedding leave and ensuring their general well-being.[37] As part of the Administrative team Banks worked closely with the Assistant Controller Valentine Goldsmith and, from 1928, with the Personnel Executive, Douglas Clarke, both of whom were responsible for general staffing issues. Clarke, in particular, oversaw the recruitment of male clerical staff.

More is known about Banks' replacement, Gweneth Freeman (although her first name was never used). Freeman initially joined the BBC in 1924, aged 24, as secretary to Goldsmith but resigned in 1927 when 'the Call O' the Wild' prompted a move to Canada (why is not known).[38] It may be that she was tempted back to the BBC in 1931 by Goldsmith, her friendship with him and her knowledge of the Administrative Department making her ideal for the supervisory role.

---

[33] *Prospero*, December 1967.

[34] Brian Hennessy (2005) *The Emergence of Broadcasting in Britain* (Lympstone: Southerleigh) p. 237.

[35] BBC/WAC:R62/100/1–3: Salary Information (staff) 1923–39. Caroline Banks, now Mrs Caroline Towler, was to return to the BBC in 1933.

[36] *The Heterodyne*, March 1931.

[37] Janet Adam Smith, who would become Assistant Editor of *The Listener*, recalled her interview with Miss Banks in 1928 describing her as 'a very, very nice person'. The Oral History of the BBC: Janet Adam Smith interview, 1 August 1979.

[38] *The Heterodyne*, April 1931.

In 1933, the post was re-designated Women's Staff Administrator as part of Reith's restructuring of the BBC. However, the new title did not confer higher status, rather it formalised Freeman's position in the administrative hierarchy. The position was well paid. In April 1939 Freeman's salary was recorded as £720 per annum, making her one of the highest earning women in the BBC.

In terms of BBC hierarchy, Freeman was arguably the most powerful woman on the staff because her decisions had repercussions for hundreds of BBC employees (630 women came within her domain in 1939). However, the role was never designated 'Director' probably because of its association with women's work. Yet despite the lack of executive standing, she worked closely with every manager in the BBC whose team included female secretarial or clerical staff, in effect most offices. All concerns about new postings, transfers, competences, ill-health, holiday leave and so on, were discussed with the WSA. Freeman was also consulted on general issues that affected women house staff and monthly paid staff, for example, she sat on the appointment boards for female salaried posts and was party to discussions on charwomen's maternity pay. She was also vital to the workings of the BBC marriage bar, as Chapter 4 will show.

Possibly her most pertinent role, however, in terms of those who came under her control, was her centrality in deciding where women would be placed. One only has to read *Ariel* and the brief pen portraits of the scores of shorthand typists and secretaries who inhabited the myriad of BBC offices to find constant references to previous bosses. This expediency had existed from the start. Ruth Cockerton, who joined the BBC in 1924, recalled telling Miss Banks that when she was given a definite job 'I hoped it would *not* be in the Engineering Department, as I hadn't got an engineering turn of mind and didn't understand what they were talking about'. Directly after, she was appointed secretary to Harold Bishop, then the Assistant Chief Engineer. Cockerton added that she had never held it against him (or Miss Banks) and that she had 'enjoyed my two years as an engineer'.[39] Secretarial staff, however, were not completely without agency. For instance, Miss Peacock, a shorthand typist with the new Television service, was given the choice of moving to Alexandra Palace or transferring to another department and remaining at Broadcasting House.[40]

---

[39] *Ariel*, June 1938.
[40] BBC/WAC:R13/436: Television: Women Staff: 1a, Peacock to Freeman, 12 August 1936.

In 1935 the *Daily Express* portrayed Freeman getting to grips with the problem of how to assist those who were moving to the new Television offices in North London, as she poured over train and bus time-tables to offer advice on their journeys. Here also was described the twice -yearly gathering in the Council Chamber for all female secretarial and clerical staff at Head Office to voice their difficulties, grumbles or suggestions.[41] As well as caring for her BBC 'girls', Freeman was outward looking and made links with her counterparts in America and in other UK-based institutions such as Miss Stretton at the Bank of England and Olga Collett at ICI.[42] It was very probably Freeman who encouraged the BBC to become a member of the Women's Employment Federation in 1935 and she developed a good working relationship with the Federation's Secretary, Ray Strachey.[43] These contacts suggest that a degree of informed cooperation existed amongst high-status administrative women.

The role of WSA encapsulated the sexual division of labour at the BBC and the belief that women had specific employment needs, necessitating a distinct female line of command. At no point was the role questioned, unlike at the Civil Service. When Hilda Martindale reluctantly assumed the role of Director of Women Establishments in 1933, she made clear her belief that 'men's and women's establishment problems were not as different as to require separate treatment'.[44] Admittedly, Martindale had notional responsibility for 77,000 women, as opposed to the hundreds in Freeman's charge, however, it demonstrates that there was a divergence of views about the need to differentiate between men and women's work. While the BBC adopted many modern employment practices, it maintained the fundamental principle that, for waged women, the control of a female administrator was essential.

The WSA's base was the General Office. It is not known if Caroline Banks had separate accommodation at Savoy Hill, but it is likely that, as female staff numbers rose, she merited a designated room to deal with confidential issues. In Broadcasting House, the General Office was purpose-built and here Freeman could work in private. *Good Housekeeping* was

---

[41] *Daily Express*, 31 August 1935.

[42] BBC/WAC:R49/371/2: Married Women Policy: 2, Freeman to Pym, 2 November 1937; 13 July 1937.

[43] Women's Library/WEF Archive:6/WEF/487: Executive Minutes, 9 May 1935.

[44] Hilda Martindale was a former Deputy Chief Inspector of Factories. Hilda Martindale (1944) *From One Generation to Another 1839–1944: A Book of Memoirs* (London: George Allen and Unwin) p. 188.

evidently enchanted by her office and how 'with its green carpet, cream walls and curtains, [it] makes such a pleasing setting for her own personal charm, of which an attractive smile, fair hair and a lovely complexion are not the least pleasing elements'.[45] Freeman also benefitted from a personal secretary and an immediate staff of three; an Assistant, a Junior Assistant and a General Office Supervisor. All were based in the 'GO'.

3.3   THE GENERAL OFFICE

The General Office (GO) was the women's hub of the BBC. Most new recruits started here, a custom dating back to the early days of Savoy Hill. In many ways it was a 'typing pool' although the term was not widely used at this time. The housing of secretarial staff in one place was a response to Taylorite notions of time-saving and efficiency that were becoming commonplace in the 1920s and 30s.[46] The journalist and novelist Ethel Mannin, who worked as a shorthand typist in London just prior to the 1920s, described the horror of the 'American system of one large general office' where one had to be on one's best behaviour all day and gossip was not allowed.[47] The General Office at the BBC was run on similarly strict lines, although the surroundings were agreeable. A description from 1938 evoked a room of about 50 girls, their desks divided into two groups of orderly rows. Careful consideration had been given to the environment under which the typists worked. The ceiling was specially soundproofed, complaints of bad lighting or faulty ventilation were few and a break was allowed for morning coffee with afternoon tea served at desks.[48] This was very different to the overcrowded, badly lit and often damp offices that many women toiled in at this time.[49]

The BBC's General Office was a means of ensuring maximum effectiveness, the logic of accommodating all the BBC's typists together was explained at Control Board in 1926. Here, the rule 'insisting upon all typists being in the General Office' was maintained 'in order to avoid waste of typists' time when non-employed in their particular section', although occasional exceptions were allowed.[50] The convention that large numbers

---

[45] *Good Housekeeping*, August 1935.

[46] Taylorism was a scientific management theory developed in the late nineteenth century whereby efficiency was improved by breaking jobs down into small tasks. Fiona McNally (1979) *Women for Hire: A Study of the Female Office Worker* (London: Macmillan) p. 79.

[47] Ethel Mannin (1932) *All Experience* (London: Jarrolds) p. 64.

[48] Wilfred Goatman (1938) *By-Ways of the BBC* (London: P.S. King) p. 25.

[49] See, for example, Joan Beauchamp (1937) *Women Who Work* (London: Lawrence and Wishart) pp. 58, 61.

[50] BBC/WAC:R3/3/2: Control Board Minutes, 30 June 1926.

of typists should be based in the 'GO' continued throughout the interwar years, often a point of friction between Freeman and an aggrieved manager. By the 1930s, those based in the General Office were copy typist. Duties included typing talks, play scripts and news bulletins and addressing the thousands of envelopes needed, for example, to acknowledge applications for jobs. The office also provided a pool of secretarial cover when an 'isolated' girl was temporarily away. These were the shorthand typists (and occasionally copy typist) who worked for individual managers or in departmental offices and who shared the same or adjacent rooms.[51]

The General Office also crucially provided a means of assessing new staff. Many women spent their three-month probationary period here during which time their skills and aptitudes were observed.[52] The WSA would then decide whether an individual should remain in the General Office or be directed to a new role. Importantly, it was where new employees were inducted into the specific practices of the BBC. Uniformity between departments and divisions was essential to the smooth running of the BBC; by learning set secretarial procedures, these could then be applied throughout the organisation. Certainly by 1937 all new recruits were issued with the 40-page manual 'Instructions to Women Clerical Staff'. The 'Instructions' including meticulous details of, for example, where to use full stops and commas, how typewriter ribbons should be requisitioned and the procedure for posting letters, as well as information on punctuality, raffles (they were not allowed) and which lavatories all but the most senior women staff were obliged to use.

Hilary Cope Morgan (who would rise to be a senior manager at *Radio Times*), began her BBC career in the General Office, an experience she described as 'hell!'[53] She recalled being asked to type news scripts using eleven carbon papers, something she had never done before. At first she was 'hopeless', managing only two sheets at a time but she soon learned the BBC way. Dorothy Singer remembered very little of her first day except that she was directed to the 'GO':

...we were taught about the BBC and the charter and impartiality and the ethos generally, you see. And then we were farmed out to anybody who wanted extra secretarial help, to various departments, and quite early on I was sent up to the D.G.'s office, and John Reith said 'keep her'. So that was how I started.[54]

[51] Goatman, *By-Ways of the BBC*, p. 26.
[52] All staff underwent a three-month trial period, whether waged or salaried.
[53] Interview with Hilary Cope Morgan.
[54] Interview with Dorothy Torry.

In 1947, Miss Singer would be appointed personal secretary to William Haley, Director General of the BBC.

3.4   THE BBC 'SECRETARY'

Maurice Gorham, the Editor of *Radio Times* from 1926, was unequivocal about the importance of the BBC secretary. When seeking information for forthcoming programmes, he professed, he preferred to liaise with the secretary rather than her producer. 'They were clever ... did the work of two men ... not only was it more pleasant, but she would probably know more too.'[55] *Radio Pictorial* left no doubt as to the significance of the 'twenty silent women', the producers' secretaries, upon whose 'unobtrusive efficiency' BBC drama and variety productions depended. 'She must be able to spot any letter which might promise something original ... she must be able to keep a secret ... she must know how to handle artistes ... she must be able to discover the names of composers and publishers...'[56] Secretaries also prompted effusive tributes in the memoirs of BBC men. Roger Eckersley, who ran the Programmes Department from 1924 before becoming Director of Entertainment, gushed about his 'perfect secretary' Miss Jockel, personally thanking her 'for the crises through which she has held my hand, for her tact with unwanted visitors, and for her technique in rescuing me from callers who would not leave'.[57] Robert Silvey, who established BBC Listener Research in 1936, was certain that 'no one could have been more loyal, conscientious or efficient' than his secretary Miss Press.[58] Hilda Matheson frequently voiced her reliance on her secretary Miss Barry, marvelling at her perspicacity, welcoming her providential return after a bout of illness and praising her ability for 'taking things in hand and calming [me] down'.[59] John Reith would end *Broadcasting over Britain* with a dedication to his secretary Miss Shields.[60]

The position of personal secretary in the interwar years was highly prized, a very different role to that of a typist. Originally a job for men (hence the

[55] Maurice Gorham (1948) *Sound and Fury* (London: Percival Marshall) p. 23.
[56] *Radio Pictorial*, 27 November 1936.
[57] Roger Eckersley (1946) *The BBC and All That* (London: Sampson Low, Marston) p. 135.
[58] Robert Silvey (1974) *Who's Listening? The Story of BBC Audience Research* (London: George Allen and Unwin) p. 23.
[59] Hilda Matheson Letters (hereafter HML), 12 February 1929, 15 January 1929, 4 January 1929, 27 January 1929.
[60] John Reith (1924) *Broadcast over Britain* (London: Hodder and Stoughton) p. 231.

continuation of the term 'Secretary' in Government departments and the Civil Service) as early as 1909 there were female private secretaries earning more than £200 a year.[61] By the 1930s the average salary for a 'good private secretary', according to J.A.R. Cairns in *Careers for Girls,* was between £5 and £10 a week although it was 'usually at the lower end'.[62] The job was the pinnacle of the secretarial promotional chain. As the private secretary Daisy Lansbury advised, it was essential not to hanker after a career of one's own.[63] In 1935, the *Radio Times* women's page included an article by Rose Rosenberg CBE, personal private secretary of 12 years standing to the former prime minister, Ramsay MacDonald. Here she described the instinctive skills for the job as 'assurance, dignity, powers of conversation, mental alertness, a real sense of value, a quick perception of people's reactions, and ... a sense of humour, which is so frequently a saving grace'.[64] The status of the secretary was always intrinsically linked to the status of her boss, indeed, to warrant a personal secretary was an emblem of prestige.[65]

At the BBC, apart from the handful who were salaried, personal secretaries rarely earned more than £5 a week. But the work was as varied as the individual she worked for: seeking out facts and information, arranging meetings, dealing with correspondence, prioritising workloads, above all understanding the BBC, her department, her boss and their different whims and needs. The relationship was often an intense one. Roger Eckersley described how Miss Jockel had stayed with him 'through fair or foul, thick or thin, for twenty long years' gratified that their 'partnership or "marriage" as master and secretary lasted longer than any other like combination in the Corporation'.[66] While secretaries at the BBC in the 1920s

---

[61] Anderson, *The White Blouse Revolution*, p. 9.

[62] J.A.R. Cairns (1928) *Careers for Girls* (London: Hutchinson and Co).

[63] Margaret Cole, ed. (1936) *The Road to Success: 20 Essays on the Choice of Career for Women* (London: Methuen) p. 77.

[64] *Radio Times*, 15 November 1935, 'The Making of a Private Secretary'.

[65] Anderson, *The White Blouse Revolution*, pp. 21–2; Davy, 'Shorthand Typists in London', p. 133.

[66] Eckersley, *The BBC and All That*, p. 57. Eckersley's reference to his 'marriage' is pertinent, later studies into secretary–boss relationships openly explore this metaphor. Kathleen Benet described the secretary as the 'substitute wife', carrying out what she defined as household duties in the office. Kathleen Benet (1972) *Secretary* (London: Sidgwick and Jackson) p. 70. Kanter, in her 1977 study of large American corporations, used the term 'office wife', stressing also the serious emotional bonds that could develop between the secretary and her boss. Rosabeth Moss Kanter (1977) *Men and Women of the Corporation* (New York: Basic Books) p. 89.

and 30s displayed some facets of 'office wife' there were also maternal elements to their interactions, clearly seen in a series of *Radio Pictorial* interviews with BBC secretaries in 1936.

For six consecutive weeks the daily working lives of seven secretaries were candidly revealed (two were joint secretaries), the motherly relationship with their managers manifest. Dorothy Knight described her boss, Eric Maschwitz, the Variety Director as being 'utterly careless of himself in every way'. She told of her horror at realising she had let him leave the office on a cold day without his heavy Ulster coat, his scarf and his hat.[67] Pamela Argent was secretary to two BBC variety producers. One of them, Bryan Michie, was an 'uncared-for-bachelor'. If she did not remind him, she claimed, 'he would always forget to send his things to the laundry and I make him swear solemnly that he will not put on anything not previously aired...'.[68] Cynthia Pughe undertook to be the memory of Val Gielgud, the Director of Drama, even phoning him at home if he needed to be in the office early, to make sure he was awake.[69] Dorothy Harrison and Ada Julian described themselves as 'the girls who "direct" Henry Hall and tell him whether he is coming or going. If we didn't he frequently wouldn't know.'[70] Henry Hall headed the BBC Dance Orchestra. These accounts provide an intriguing counterpoint to the heady creativeness of these senior officials of the Variety and Drama Departments, as evoked by Paddy Scannell and David Cardiff in their social history of the BBC.[71]

Matheson's relationship with her secretary appears to have been more personal than was the case with a male manager. For instance, she was deeply concerned for Eileen Barry's welfare when her relationship with Derek McCulloch (then a BBC announcer) was jeopardised by his transfer to Belfast.[72] Barry also appears to have been aware, and supportive, of Matheson's relationship with Vita Sackville-West, 'beaming' when asked how she had guessed that Matheson was staying at Sissinghurst and going

---

[67] *Radio Pictorial*, 17 January 1936.

[68] *Radio Pictorial*, 14 February 1936.

[69] *Radio Pictorial*, 7 February 1936.

[70] *Radio Pictorial*, 31 January 1936. Henry Hall was one of the UK's foremost bandleaders.

[71] Paddy Scannell and David Cardiff (1991) *A Social History of British Broadcasting, 1922–1939* (London: Basil Blackwood) pp. 256–73.

[72] HML, 22 February 1929. McCulloch would become Director of Children's Hour. The couple married in 1931 at which point Barry left the BBC.

'quite pink' when Vita personally sent her lupins.[73] Whereas Matheson, in her letters, was only ever highly complimentary about Barry she was less than polite about 'the DG's secretary', Elizabeth Nash, describing her as an 'uppish young woman with an exaggerated sense of her own importance'.[74] As a rare BBC secretary who was on the salaried staff it is perhaps not surprising that Miss Nash considered herself superior. In 1928, when she took over the 'No. 1' job she earned £300, when she left in 1936, this had risen to £600 a year.

Nash was at the pinnacle of the BBC's secretarial chain. In 1935, at the insistence of the new Controller of Programmes Cecil Graves (largely to benefit his own secretary Margaret Hope Simpson) it was agreed that the four Controllers' secretaries would also be salaried.[75] For a BBC woman whose ambition was a secretarial career it was anticipated that, if she remained on the staff and performed well, she would rise up the hierarchical chain from copy typist or shorthand typist, to 'isolated' shorthand typist and ultimately, secretary. This expectation was not limited to the BBC. Writing in *Careers for Girls* in 1928, the 'secretary to the editor of a leading London newspaper', observed that once beyond the role of 'machine' (albeit a superb one, the writer stressed, with speed and accuracy that made her a reliable and valuable typist) 'all sorts of horizons became visible'.[76]

The designation 'secretary' is significant. For many years there was incongruity within the BBC's secretarial ranks which typifies its complex and bizarre hierarchical structure. Until 1937 only a woman who worked for a senior executive was officially a 'secretary'. A shorthand typist might effectively be a secretary but if her boss was not high-ranking she could not be referred to as such. This is clearly seen in the 1937 Staff List. The top official in each department has a personal secretary; senior managers a designated shorthand typist; while for those in more junior management positions, the shorthand typist might be shared. Freeman freely admitted that the designation was a misnomer as 'all these girls are referred to— quite naturally—as secretaries'.[77] The situation was finally rectified in that

[73] HML, 15 January 1929, 27 May 1929. Sissinghurst was Vita's Kent home.
[74] HML, 5 January 1929.
[75] BBC/WAC:L1/1699/1: Margaret Hope Simpson Staff File. Miss George, Miss Osborne and Miss Shawyer were the other three.
[76] Cairns, *Careers for Girls*, p. 30, 39.
[77] BBC/WAC:R49/607: Salary Review: Women Staff, Freeman to Clarke, 6 December 1937.

year's Salary Review which clarified that all 'isolated' shorthand typists would be redesignate secretaries, a truer reflection of their role.[78] It was made clear, however, that while there might be a change of name, it did not bring with it any rise in pay. Rather, it was hoped the new designation would 'do away with the present dissatisfaction caused by the existing "hierarchy" system'. Salaried secretaries were in future to be called 'senior secretaries'.[79]

The secretary–boss relationship has long been associated with sex, indeed Maurice Gorham claimed of the BBC that he 'never knew an office where sex played so large a part, where so many people lived with their secretaries, where the hunters and the hunted were so conspicuous as they went about their sport'.[80] He was convinced that while people dressed respectably they did not behave respectably. There were certainly marriages between secretarial staff and senior BBC officials. Valentine Goldsmith, the Head of Administration was 'ragged' at Control Board about his impending wedding to Miss Fawkes in June 1927 who Reith described in his diaries as 'one of the junior typists, rather well-known for her irresponsibility'.[81] Although couched in ambiguity, it seems almost certain that the marriage, which took place in September, was prompted by Fawkes' pregnancy. The tragedy of her death shortly afterwards from 'electrical treatment to stop this' and Goldsmith's evident distress (according to Reith he had known nothing about it) was 'a dreadful business', painfully recorded in the diary entry for 22 November 1927.[82] Goldsmith's marriage was evidently harrowing. Others were more joyful affairs as was the case with, for example, Eileen Barry and Derek McCulloch, who married in 1931.

While the BBC's secretarial staff were mostly female, there were times when a male typist was employed. For example, it was seen as prudent in the case of Edward Clark, the Music Director at the Newcastle Station, whose language at times of stress was 'most unpleasant'.[83] Appropriateness

---

[78] BBC/WAC:R3/3/12: Control Board Minutes, 21 December 1937.

[79] Salary Review, Freeman to Clarke, 6 December 1937.

[80] Gorham, *Sound and Fury*, p. 20.

[81] Reith Diaries, 10 June, 21 June 1927. Reith described Goldsmith as 'much embarrassed and obviously wished she had been in a different position'.

[82] Reith Diaries, 22 November 1927. The story was covered in the press, but Goldsmith was not named.

[83] As quoted in Charlotte Higgins (2015) *This New Noise: The Extraordinary Birth and Troubled Life of the BBC* (London: Guardian Books) p. 45.

was also probably the reason why, as the expansion of the BBC necessitated night work, the two typists employed in the General Office were both men.[84] The BBC was not alone in employing small numbers of male typists. Hilda Martindale noted that it might be necessary for the Civil Service to employ men in this capacity 'in a few special departmental situations' with all male staffs.[85] This was mirrored at the BBC. While male typists were occasionally employed in Regional offices, most were employed in areas of work that were predominantly, or exclusively, male. These were the Advertising Department, the Engineering Equipment Department and at the BBC's nine transmitting stations. By definition these stations were in isolated parts of the country so it was seen as especially inappropriate to employ women and even the telephonists and cleaners were male.[86]

3.5   The Telephone Exchange and Duplicating Office

As exemplified by the transmitting station, communications were at the heart of the BBC. Telephones and duplicating machines were similarly emblematic of this and were also a symbol of the modern twentieth-century office. At the BBC it was women who operated these technologies, in line with the convention which saw routine office work as an appropriate female occupation.[87] The telephone and duplicating machine feature in the recollections of BBC staff. Even at Magnet House, Cecil Lewis marvelled how they never stopped.[88] Ruth Cockerton, looking back to Savoy Hill in the mid-1920s, recalled the Roneo machine 'which ground out memos and things in a room where people hung their hats and coats'.[89] The operator of this lone machine was a 'girl'. By 1938, the Duplicating Section at Broadcasting House employed 27 female operatives.[90] The BBC telephone exchange witnessed a similar expansion. From one operator in 1923, by 1938 it had a staff of 25. Since the first exchanges had begun

---

[84] BBC Staff List, 1937. Protectionist legislation forbade the night employment of women in many industries and the BBC may have been influenced by these laws.

[85] Hilda Martindale (1938) *Women Servants of the State, 1870–1938* (London: Allen and Unwin) p. 91.

[86] The exception was Bournemouth which employed a female cleaner.

[87] Anderson, *The White Blouse Revolution*, pp. 17–18.

[88] Cecil Lewis (1924) *Broadcasting from Within* (London: George Newnes) p. 27.

[89] *Ariel*, June 1938.

[90] Goatman, *By-Ways of the BBC*, p. 29.

operating in the 1880s, telephonists had been women. By 1921, the Post Office employed 26,000 female telephonists and the GPO would become the main recruiting ground for the BBC.[91]

Amongst Reith's first appointments in 1923 was Olive May, recruited specifically to operate the Private Branch Exchange. She was personally interviewed by Reith who stressed to her the importance of the job. The switchboard was the first point of contact between the outsider and the BBC he emphasised, and as such, it was vital that all callers were dealt with intelligently, efficiently and courteously.[92] However, the space was cramped, the hours of work long (including weekends), with no overtime pay. Miss May was soon joined by a second switchboard operator and together they covered extensive shifts, starting at 9 am and finishing late, often not until 10.30 pm. Despite the intensity of the work, May relished her time at Savoy Hill, especially the amity that developed between her and Reith. In 1928 she married a BBC engineer from Leeds, whom she had met over the phone, and she subsequently resigned. However in 1931, now Mrs Bottle, she returned to the BBC for twelve months, to oversee the development of the telephone exchange at Broadcasting House.

The new switchboard, when it opened in 1932, had six telephonist positions. By 1938, this had been extended to 12.[93] Not only did the 23 operators, working in shifts, deal with an average of 11,000 calls each day but their job also entailed memorising the details of 650 internal extensions and the frequent movements of staff. A photograph from this time shows a row of women telephonists dressed in neat white blouses and dark skirts, their female supervisor, in a smart floral dress, standing over them.[94] Reith maintained a close relationship with the switchboard. Dorothy Torry, who worked in his office from 1936, remembered how a senior telephonist was employed to deal personally with Reith and how he made a friend of her. 'They knew each other and their wants and dislikes and the way they spoke. He felt he couldn't really manage with someone different', she explained.[95] This sense of a personal connection with the BBC's telephonists extended to other senior staff. In 1946,

[91] McNally, *Women for Hire*, p. 5. The use of women as GPO telephonists was a legacy of telegraphy, where the experiment of employing female staff had first taken place. Martindale, *Women Servants of the State*, pp. 16–17.

[92] Sound Archive No: 87181, 'Reith Remembered', broadcast 21 June 1989.

[93] Goatman, *By-Ways of the BBC*, pp. 32–3.

[94] Goatman, p. 33.

[95] Dorothy Torry interview.

Roger Eckersley looked back nostalgically to earlier days when 'the girls in our telephone exchange used to know us all—I still feel hurt when a girl asks me to repeat my name'.[96] At first the telephone exchange epitomised the familial ethos of the early BBC. Later it became its corporate voice.

According to *Radio Pictorial* in September 1939, the 'Hello Girls' of the BBC needed long-term experience in a Post Office exchange before they could be considered for a position with the Corporation.[97] As with other secretarial and clerical staff, the BBC expected telephonists to arrive already trained. Four years' London Telephone Service (LTS) experience was required and while languages were not essential, they were an asset. The average starting age for a BBC telephonist was 24 on a rate of £3 a week, a rate that compared favourably with other employers such as Harrods, Midland Bank and Selfridges.[98] However, it was not only the pay that made the BBC an attractive place to work, it was the prestige of the job. As Reith had originally told Miss May, switchboard operators dealt with people from all walks of life and the nature of BBC work would have entailed daily contact with dignitaries and celebrities as well as the public at large.

For those who worked in the Duplicating Section, daily life was less glamorous, but the busy office was a vital hub of the BBC. Wilfred Goatman, writing about the office in his 1938 guidebook about the BBC, joked that the duplication of forms was a major function of the Section, listing with relish the thousands of specially formulated index cards, internal memorandum sheets and analytical record pro formas that were printed every day. The Section Supervisor had even designed a form for the requisitioning of forms, he smirked.[99] While this may seem trivial, a large bureaucracy, such as the BBC, needed a uniformity of paperwork to function efficiently. The Duplicating Section not only produced forms, it was also responsible for numerous other tasks that included the copying of play-scripts, minutes of meetings, press releases and announcer's duty sheets along with daily menus and studio allocations, all of which might be subject to last minute change.[100] Mary Lewis, who joined the Duplicating Section as a checking clerk in 1938, described an atmosphere of friendship

[96] Eckersley, *The BBC and All That*, p. 66.

[97] *Radio Pictorial*, 8 September 1939.

[98] BBC/WAC:R49/237/1: Grades and Salaries: Telephonists, Freeman to Wade, 22 March 1937. Freeman believed telephonists should be graded BW rather than CW.

[99] Goatman, *By-Ways of the BBC*, p. 28.

[100] *Everywoman*, February 1935.

within the office, this despite the rigid discipline that was enforced by the Supervisor, Miss Hills.[101]

Duplicating is not a job singled out in studies of women's work in the interwar years, so it is difficult to compare the BBC's duplicating staff with others doing similar work. However, conditions of service were in line with all BBC weekly paid secretarial and clerical posts with operatives placed in two grades earning up to £3.10s a week.[102] According to the 1937 Staff List, under the management of a Supervisor and Assistant Supervisor, were two clerks, seven stencil typists, fourteen operators, two office girls and three 'boys'.[103] Duplicating, like telephony, was a specialist job and there is no indication that any women who worked in these sections were ever considered for promotion or transference to other areas of work. Mrs Rouse, the PBX supervisor, Miss Hills, the Duplicating Supervisor and Miss Armstrong, the Assistant Duplicating Supervisor (all salaried positions by 1939), had risen through the ranks of their particular section which was the norm in the interwar years, particularly for manual jobs.[104]

3.6   THE BBC CLERK

The role of the BBC clerk is seldom described. There are scant references in newspaper articles presumably because the work did not have the glamour or interest that typists, telephonists or even duplicating operatives did. The responsibilities would have varied hugely depending on which section of the BBC the clerk worked. For example, Lilian Taylor, briefly in Programme Finance (from 1926–28) had amongst her responsibilities the maintenance of a card index of Artists who were booked by the Music, Production and Talks Sections. She recorded details of their fees which ensured that the appropriate payment was made for each subsequent booking, regardless of which Section employed them. She also prepared the Artists Expenses Sheets, making certain they had been paid.[105] Miss Wallage, a clerk in Premises and Standing Charges, was described in

---

[101] Mary Lewis interview.

[102] Roneo and Multigraph operators were graded 'C3W' earning up to £3.10s weekly, Junior Duplicating Machine Operators were graded 'DW' earning up to £3 a week.

[103] BBC Staff List, 1937.

[104] Female factory workers could become charge-hands, forewomen or supervisors although most stayed in the same grade for their entire working life. Glucksmann, *Women Assemble*, pp. 111, 202–3.

[105] BBC/WAC:R13/305: Programme Contracts, Crutwell to Eckersley, 31 January 1927.

1937 as being personally responsible for the accounts and statistics connected with the Post Office lines used for Outside Broadcasts; work that
was 'specialised and of a somewhat unattractive type'.[106] Clare Lawson
Dick's clerical role in the Registry was to file letters while Mary Lewis's
in the Duplicating Section was to check the quality of the job. The BBC
employed both women and men as clerks; in 1934, for example, 86 male
and 68 female clerks worked for the Corporation.[107]

It is difficult precisely to compare the work of the BBC's male and
female clerks as they rarely worked side by side. However, they were
employed in the same grades and payment bands, which was unusual for
the times.[108] Women clerks worked in the Registry, in Programme Finance
and Salaries and in the Gramophone and Music Executives. Male clerks
worked chiefly in four BBC areas: Accounts, Catering, Equipment and
Publications.[109] While a small number of clerical jobs were interchange-
able, in most instances there appears to have been a specific decision to
employ either a male or female clerk. In the Advertising Department,
the Publications Department and the Post Room, all predominantly
male departments, clerks were men. In the Duplicating Department, the
Registry, Office Administration and Programme Administration, clerks
were women. Only two small departments employed equal numbers
of male and female clerks: Staff Records and Display. Staff Records was
part of the Staff Administration department which included a significant
number of senior women amongst its managers. The Display section was
headed by Kathleen Lines which suggests that while male bosses had a
preference for working with male clerks, female managers were more open
to working with a mixed staff.

There are hints that male clerks were viewed as more valuable to the
Corporation. For example, a memo from the Programme Correspondence
Section in 1934 pointed out that 'it was a waste of a man's ability' to keep
him on routine, simple work such as answering easy enquiries, copying
talks statistics and checking dance band correspondence.[110] Rather, it was

[106] BBC/WAC:R49/372: Married Women Policy: Tribunals, Tribunal held on 2 June
1937.
[107] Salary Information Files.
[108] Salary Information Files shows six women and 20 men began their BBC careers as
waged clerks. Their wage rates reveal broadly equal pay.
[109] BBC Staff Lists, 1934, 1937.
[110] BBC/WAC:R13/395: Programme Correspondence, Freeston to Programme Services
Executive, 8 June 1934.

suggested that a Miss Grieves should be taken out of the General Office and made responsible for the work 'as described as suitable for a woman'. This request was strongly supported by Nicolls, the Director of Internal Administration who commented, 'I have felt for some time that material is being wasted through the men having to devote so much time to dealing with simple enquiries and routine work which could be done so easily by someone like Miss Grieves'. Whether Miss Grieves aspired to a long-term BBC career is not known, but many women clerks took their positions very seriously, particularly if they sought to rise to the salaried ranks. This opportunity for promotion, however, was not a possibility for the BBC's House Staff.

3.7   WOMEN HOUSE STAFF

An organisation as large as the BBC required an army of service staff to maintain it, in particular offices and studios needed to be cleaned, cloak-rooms staffed and refreshments and meals prepared and served. This predominantly was work for women, their recruitment and their day-to-day working lives the responsibility of the House Superintendent, H.L. Chilman. While Savoy Hill required only a small team, by 1939, Chilman's empire at Broadcasting House encompassed many hundreds of manual staff. Jobs were gendered. Commissaries, studio attendants, kitchen porters and store staff were positions held exclusively by men. Women were employed as lavatory attendants, kitchen cashiers and counter-hands as well as waitresses and charwomen. A rare job that was open to both women and men was that of 'chef'.[111]

Reminiscing, in 1937, on her time at the Corporation, the former BBC accompanist-turned-music star, Jean Melville, posed the question:

> Do you know the two women to whom I think I was the most deeply grateful during my time at the BBC? They were Mrs Dubarry and the late Mrs Hudson—and their job was a very simple but important one. They have looked after the BBC canteen since Savoy Hill days.[112]

The BBC's restaurants and canteens were significant places for BBC staff. Not only were they a crucial part of the Corporation's welfare provision

---

[111] BBC Staff List, 1937.
[112] *Radio Pictorial*, 25 June 1937, 'BBC from the Inside' by Jean Melville.

but they were also an attractive place to meet and share news and gossip. When Wilfred Goatman toured the restaurant in 1938 for his guidebook to Broadcasting House he noted that the 86 day-staff employed in the restaurant worked three different shifts and provided thousands of meals each week.[113] He also proudly depicted the state-of-the-art kitchen with its 'four electric ovens, two large grills, pastry oven, steamer, four fryers, a large electric carver and refrigeration plant'.

In 1937, under the supervision of A.E. Mason the Head of the Catering Section, were three chefs, two second chefs, two assistant chefs, a commis chef, a patissiere and a cook-chargehand. All were women, except for one of the chefs and an additional night chef. To be a cook or a chef in interwar Britain was a respected position which would have entailed training, either at college or on-the-job. In her 1935 careers guide, Ray Strachey discussed the availability of full-time courses at Domestic Science Colleges and private cookery schools along with day and evening classes at Polytechnics and other further education colleges. Salaries for non-resident cooks, she estimated, averaged £100 to £150 per annum, with kitchen superintendents possibly earning as much as £220.[114] How the BBC's chefs and cooks were trained is not known but in 1939, Mrs Broughton, the 52-year-old Head Chef at Broadcasting House, earned the highest wage of all the catering staff, an exceptional £5.18s 9d a week, equivalent to around £300 a year.

Waitresses were the largest single group of BBC catering staff and again, their conditions of service appear to be good. Goatman wrote whimsically about the 'little room where the waitresses have their meals in comfort and are themselves waited upon by a girl employed solely for that purpose'.[115] This impression is in stark contrast to Joan Beauchamp's portrayal of the miserable working conditions of a waitress in one of the largest teashop firms in the UK (presumably J. Lyons). The work was tiring, the changing room 'an old cellar, damp and horrible and far too small' and if a girl felt ill, there was nowhere to sit down.[116] A later study into J. Lyons, the home of the iconic 'nippy', confirmed low pay and a ruthless management style which could lead to instant dismissal for flouting one of many petty rules.[117]

---

[113] Goatman, *By-Ways of the BBC*, p. 49.
[114] Strachey, *Careers and Openings*, pp. 217–18.
[115] Goatman, *By-Ways of the BBC*, p. 51.
[116] Beauchamp, *Women Who Work*, p. 49.
[117] Glucksmann, *Women Assemble*, pp. 129–30. A 1929 Inquiry into the Catering Trade confirmed that hours were long and wages low, Strachey, *Careers and Openings*, p. 112.

At the BBC a model letter of appointment, probably from 1938, shows a starting wage of £1.10s per week and free meals.[118] Discussions on whether the BBC should abide by the 1913 Shops Act (it did) clarified that waitresses were eligible to three weeks' holiday a year, two weeks more than the statutory minimum.[119] Questioned in 1934 as to whether a sufficiently good wage was being paid to attract 'a decent class of girl', Ralph Wade, the Director of Office Administration, was adamant they were; the BBC paid 30s a week as compared to the 27s 6d paid by Lyons.[120] In addition, BBC girls had meals of better quality and quantity, were provided with uniforms, including shoes and stockings, and were offered free laundry.[121] There was camaraderie too, as *Ariel* reported in June 1936:

> If your tea at Broadcasting House and Maida Vale tasted salty in the afternoon on Monday, May 18 it must have been because all the girls of the kitchen and restaurant staff had had a day out at Margate the day before, in charge of Mrs Cox and Mrs Dubarry. They filled two charabancs arriving there at midday for a blow-out in a big hotel…[122]

Kitchen staff appear to have held a higher status than waitresses. Whereas the former might be considered for the BBC pension scheme waitresses 'should not be eligible in any circumstances'.[123] This differentiation was probably linked to the high turnover of waitresses whose average length of service at the BBC was three years.

Charwomen, on the other hand, might stay with the BBC for many years, even decades. The pay was good, the hours of work fair and, attached to the job was glamour and cachet. A photograph in *Radio Pictorial* from 1934 showing the BBC's charwomen leaving Broadcasting House at the end of their shift, is captioned 'a happy band of cleaners'.[124] And they do look both happy and respectable. To be a charlady at the BBC was a sought-after position. Chilman reported in 1936 that he had 2,000 on his

---

[118] BBC/WAC:R49/73/1: Catering Staff: Conditions of Service, Catering Manager, Letter of Appointment, 1938.

[119] Catering Manager's Statement, 10 March 1938.

[120] According to Routh the average wage for a waitress was £113. Guy Routh (1965) *Occupation and Pay in Great Britain 1906–1960* (Cambridge: Cambridge University Press) p. 95.

[121] BBC/WAC:R49/73/1, Wade to Nicolls, 11 January 1934.

[122] *Ariel,* June 1936.

[123] BBC/WAC:R49/74: Catering Staff: Wages, Clarke to Wade, 20 September 1933.

[124] *Radio Pictorial,* 28 March 1934.

waiting list compiled from individual applications as well as from recommendations from Governors, members of the staff and 'all sorts of eminent people in all walks of life'. When vacancies occurred interviews were in strict order of application, he explained, at that time he was dealing with women who had applied in January 1935.[125] In 1939 there were 214 charwomen working in London and 86 in the Regions.[126]

Once in the job it was obviously one to keep, as demonstrated by longevity of service. For instance, *Ariel* in 1948 congratulated Mrs E. Simpson, a Charwoman in Edinburgh on the completion of 21 years' service and, in 1954, Mrs Mary Leonard, a Senior Cleaner in Belfast for 30 years with the BBC. The ten-year's service of the Forewoman Cleaner and her three Assistant Forewoman Cleaners at Broadcasting House was celebrated in *Ariel* in 1937. The four had started work at the BBC (it would have then been based at Savoy Hill) on the same day in November 1926, suggesting a high level of friendship.[127] A hint at the duties of the BBC's charwomen was given in a 1935 *Radio Pictorial* article which calculated that in London each day they polished 107 mirrors, filled 131 soap bottles and used 500 gallons of soap. The reporter fantasised about sauntering into Broadcasting House when the charwomen arrived at 7.30 am, watching their names being ticked off in a huge ledger and then dispersing to their duties. Some were mothers, some were grandmothers, some were war widows, he claimed, 'It does not signify. All that matters is—the power in the elbow.'[128]

Almost all the BBC's charwomen were married. At a time when married women's employment was discouraged, cleaning and charring were the most acceptable way to bring extra income into the working-class home.[129] Sylvia Anthony, writing in 1932, estimated that the normal rate for a London charwoman was 9d to 1s an hour.[130] Details of the wages

[125] BBC/WAC:R49/56/1: Recruitment of Staff:1, Chilman to Clarke, 26 November 1936.
[126] BBC/WAC:R/49/178/16: Staff Policy: Establishments: 1 July 1939.
[127] *Ariel*, June 1937.
[128] *Radio Pictorial*, 28 June 1935.
[129] Ray Strachey estimated there were 39,000 married women who earned money through casual domestic work, charring and office cleaning, Ray Strachey ed. (1936) *Our Freedom and its Results by Five Women* (London: Hogarth Press) p. 148. At the BBC, according to the BBC Staff List 1937, out of 83 charwomen employed at the Regional offices, 73 were married.
[130] Sylvia Anthony (1932) *Women's Place in Industry and Home* (London: George Routledge) p. 27.

paid to BBC's charwomen are unavailable. However, in May 1937 a Mrs G. Saunders was retired from service owing to ill health and was paid £39, purportedly a year's wages.[131] This would have made her earnings 15s a week for two-and-a-half hours' work each morning, above the rate suggested by Anthony. That charwomen were viewed differently from other female staff is clear from the deliberations on the marriage bar: they were exempt from the start because they were 'traditionally married women'.[132] In June 1938, at the discretion of the Corporation, they were granted maternity leave including an ex gratia payment equivalent to four weeks' wages.[133] The practice was formalised in 1939 with eligibility being clarified as a minimum of one year's service, a good report and the declared intention of continuing with the Corporation after the birth.[134] There were also moves at this time to admit charwomen to the BBC pension scheme.

The charwomen's smart appearance and the orderliness of the waitress in her laundered uniform were in line with an organisation that wanted to project probity and decorum. The BBC was keen to employ reputable staff. Although the status of house staff was lower than that of the BBC's office-based employees, those with long service were viewed as part of the BBC family, for instance their photographs and signatures were included in the book commemorating Ten Years' Service which would have involved a congratulatory meeting with Reith. Yet house staff, presumably because of their association with manual labour, were never eligible for salaried status. Even those in supervisory roles whose pay might rise to be the equivalent of £260 a year or more were denied this reward. To be salaried was a privilege open only to office-based staff, and only then to those deemed suitable.[135]

3.8   Promotion to the Salaried Grades

In 1937 Beatrice Hart, a weekly paid secretary in the Supplementary Publications Department, was recommended by her line-manager for promotion to the monthly paid grades. This was endorsed by the head of

---

[131] BBC/WAC:R3/3/12: Control Board Minutes, 1937.

[132] BBC/WAC:R49/371/1: Married Women Policy: 1, Goldsmith to All Regional Directors, 20 October 1932.

[133] BBC/WAC:R49/326/1: Maternity and Grants, Clarke to Regional Directors, 7 June 1938.

[134] Pym to Nicolls, 19 July 1939.

[135] The BBC's salaried engineers were considered separately.

department, who agreed that 'the stage has been reached when Miss Hart should be graded as a junior assistant'.[136] The possibility of promotion to a salaried position, where the conditions of service were far superior, was an important aspect of work at the BBC and set the Corporation apart from many office-based jobs in the interwar years which could be notoriously dead-end. The Civil Service was also seen as a good place for advancement. Young women who joined the Manipulative Grades as weekly paid Writing Assistants or as Typists were eligible to enter the examination for the salaried Clerical Grade. However, the competitive nature of the examination, the over-achievement of men and the many years it took to reach an appropriate level to apply for promotion to the Clerical Grades meant this could take many years. GPO records show for some women it took 20 years or more to gain promotion.[137]

Opportunities for self-advancement were a crucial element of the work ethic in the 1920s and 30s and highly pertinent in British society where social class was still so important.[138] The chance to 'get on' was always a focus of career guides, where 'character' and 'personality' were seen to be as critical as ambition. Dorothy Hughes was certain that while nearly all occupational doors were now open to women 'ability and character decide in each case how wide open and how far the individual will go'.[139] This was echoed by Vrywny Biscoe who rejoiced that the 'infinite variety of opportunities at hand' for the college girl were also available to the non-college girl 'given, of course, that she has that in her which makes for "character"'.[140] How much 'character' a young woman might have depended largely on her upbringing and education; a 14-year-old elementary school leaver would always be at a disadvantage to those who remained at school until 16 or 18.[141] The chance of promotion, though, was mostly dependent on the type of workplace one was in. However much 'character' and ambition a young woman might have, if there was no mechanism for advancement, she would find herself stuck.

At the BBC, for the brightest and most ambitious waged women, those who from the start showed themselves to be capable of greater

[136] BBC/WAC:L1/1698/1 Beatrice Hart staff file, Confidential Report, 1937.
[137] Dorothy Evans (1934) *Women and the Civil Service* (London: Pitman) pp. 81–93.
[138] Ross McKibbin (1998) *Classes and Cultures: England 1918–1951* (Oxford: Oxford University Press) pp. 44–50.
[139] D.W. Hughes (1936) *Careers for our Daughters* (London: A&C Black) p. vi.
[140] Vyrnwy Biscoe (1932) *300 Careers for Women* (London: Lovat Dickson) p. 7.
[141] McKibbin, *Classes and Cultures*, pp. 259–71.

responsibilities, the route to the salaried grades could be quick and easy. For others, as was the case in the Civil Service, the process could be painfully slow. It took Beatrice Hart ten years to reach the salaried grades; for Alice Wright in the Music Library the process took 14. In the case of Miss Wright, her personal file shows that the problem lay not with her direct manager, the Music Librarian, who for many years advocated her promotion to the salaried grades but with his seniors.[142] This lack of uniformity and transparency was endemic at the BBC. Promotion and mobility were largely at the whim of the manager, or managers, concerned. Clare Lawson Dick, for example, was convinced that Miss Freeman looked upon those who tried to move upwards as 'conceited or even rebellious'. She recalled being summoned to the WSA's office to be told that she had 'insulted' her seniors by applying for the position of play-reader in the Drama Department in the incorrect manner, she should have sought permission from Freeman first (Lawson Dick had performed well at the interview but did not get the post).[143] Although this vindictive attitude is not borne out in the extant personal files of those who did gain promotion to salaried positions, where Freeman is shown to be supportive, it is eminently possible that she held back those she disliked. Lawson Dick also told the story of Josephine Plummer (who rose to be Assistant Head of *Children's Hour*) who saw a note in her file written by Freeman which read 'will need to be ridden with a tight rein'.

Aside from those who applied for a monthly paid post or who were plucked out for one, it is unclear how, when, or why weekly paid individuals were considered suitable for salaried status jobs within the BBC. For some, the fact of reaching £5 a week, the 'roof' for most weekly waged positions (and the equivalent of £260 a year), seems to have prompted a redesignation of role, with an accompanying move to the salaried grades. For others whose wage rose to £5 a week, there might be a switch to monthly pay but with no change of title. Still others who earned £5 remained on weekly wages, often for long periods, as was the case with Alice Wright. It is not known how many other women continued on their £5 'roof' without ever gaining a salaried position, as only those who were promoted to the monthly paid grades were recorded in the Salary Information Files.

[142] BBC/WAC:L1/7291/1: Alice Wright Staff File, Confidential Reports, 1933, 1935.
[143] Clare Lawson Dick interview.

There were a number of 'rags-to-riches' success stories at the BBC, as mirrored in the popular books and films of the day. To have a meteoric career rise was not unheard of in the 1920s and 30s. As Elsie Lang averred in her book *British Women in the Twentieth Century*, 'the majority of the great business women today have climbed into positions of £1,000 a year and even more by means of shorthand and typing'.[144] The wealthy industrialist Lady Rhondda advised aspiring young businesswomen to start as shorthand typists in an environment where there was no prejudice against women and where, by making herself useful, she might rise to a position of power.[145] In her 1936 guide *Careers for our Daughters*, Dorothy Hughes enthused about the university and secondary school girls who had 'qualified for executive responsibility' at the BBC pinpointing three women who had started as secretaries and others who used their training as musicians or as a drama student to move into lucrative and important positions within the Corporation. Almost certainly one of the 'musicians' was Doris Arnold, a BBC shorthand typist who became a variety star. Her BBC career and that of Mary Hope Allen, who rose from cataloguer to drama producer, are considered in Chapter 5.

Another example of a young woman pushing for change in her career is Joan Vickery. In 1935 Vickery, a shorthand typist and then a clerk in Outside Broadcasts, was keen to transfer to the fledgling Television Department. Freeman was supportive, aware that the Department would soon need 'girls of her calibre' and keen that she should be in 'on the ground floor'.[146] Vickery's manager, Gerald Beadle echoed Freeman's opinion, clear that he would not 'stand in the way of Miss Vickery's advancement, because I think she deserves it'.[147] Vickery's transfer was agreed and within two years, she had been promoted to the salaried grades. Other weekly paid secretarial and clerical women who moved to significant posts include Winifred Baker, who ran the Manchester Orchestra; Marjorie Redman who moved from a typist in Education to Sub-Editor

---

[144] Elsie M. Lang (1929) *British Women in the Twentieth Century* (London: T. Werner Laurie) p. 259.

[145] Lady Rhondda in *Ladies Field*, quoted in Angela V. John (2013) *Turning the Tide: The Life of Lady Rhondda* (Cardigan: Parthian) pp. 277–8.

[146] BBC/WAC:R13/426/1: Television Department: Women Staff, Freeman to Clarke, 10 May 1935.

[147] Beadle to Clarke, 14 May 1935. In fact, Beadle requested that Miss Vickery's departure be delayed because of staffing shortages in his department. This was over-ruled.

of *The Listener* and Evelyn Shepherd who, over a period of fourteen years, was promoted from a £1 a week clerk to a £320 a year Furniture Buyer.

There was a clear difference in the pace at which men and women were promoted to the salaried grades, men generally moved up the ladder far more quickly.[148] It is also clear that far more senior women than men began their BBC careers as weekly paid, 53 per cent of women compared to 8.5 per cent of men. A major reason is that far fewer men were employed in secretarial or clerical positions (the majority of waged male staff were engineers) but another consideration is that the BBC was less prepared to recruit women directly to 'officer class' posts as Chapter 5 will show. In the 1930s there was still anxiety among feminists that it was brothers, rather than sisters, who were given opportunities to go to university.[149] At the BBC it was the graduate brothers who were recruited directly to the salaried staff; their non-graduate sisters might have to be content, initially, with the weekly paid grades. A small but significant number of female graduates also joined the BBC this way.

The President of the Women's Employment Federation, Grace Hadow, believed starting as waged could be a positive choice, claiming 'not all university girls who take jobs as shorthand typists will have to spend the rest of their lives taking down letters. I know of many girls who deliberately seek these jobs as a stepping-stone to something better.'[150] In November 1935, Freeman reported that eight 'university girls' were employed by the BBC in Registry work.[151] The following year *Ariel* informed its readers that eight of Edinburgh's secretarial staff were graduates.[152] Isa Benzie, who became Foreign Director, and Cecelia Reeves, who went on to be Paris Representative, both started their BBC careers as secretaries. Elizabeth Barker, who would become Head of European Talks and English Service, had been at Oxford before joining the BBC as a clerk in 1934 while Mary Lewis, a graduate of Westfield College, started as a duplicating clerk in 1938. Her BBC career saw her rise to be Head of Pay Policy.

---

[148] Salary Information Files.

[149] See, for example, Mary Agnes Hamilton in Strachey, *Our Freedom*, pp. 253–4; Vera Brittain in Paul Berry and Alan Bishop, eds. (1985) *Testament of a Generation: The Journalism of Vera Brittain and Winifred Holtby* (London: Virago) pp. 121–2, 126. As late as 1939, only 23 per cent of university students were female, Carol Dyhouse (1995) *No Distinction of Sex? Women in British Universities, 1870–1939* (London: UCL Press) p. 7.

[150] Press Association interview, 25 January 1934.

[151] BBC/WAC:R13/399: Registry Staff, Freeman to Reith, 12 November 1935.

[152] *Ariel*, June 1936.

In 1934, an independent 'Report into the Recruitment of Staff' raised concerns about Cambridge and Oxford Universities being approached with regards to graduate recruitment to weekly paid posts, as these were 'candidates whose natural entry is to the monthly-paid staff'.[153] The fear was that an increased number of university women in waged jobs 'might deprive the other weekly-paid women of all chance of promotion to the monthly-paid staff'. This was also a concern of Ray Strachey who advised that the most prized secretarial posts tended to go to those who had gone to university.[154] Miss Freeman, however, was adamant that this was not the case at the BBC as the 'fact that a girl has a degree' was a very secondary consideration; she had to have 'many more qualifications besides'.[155] Nine out of ten recruits from universities were quite unfitted for secretarial work, Freeman claimed, but it was 'worth interviewing the ten to get the one'. Clare Lawson Dick's assertion that well-educated girls or graduates could be seen as problematic is evident here. But Lawson Dick did get a BBC job, and did get a promotion; persistence and ambition patently could pay off.

3.9   CONCLUSION

When Dorothy Singer joined the BBC as a shorthand typist in 1936 she was thrilled to start work on £2.10s a week which she declared was 'rather large and a lot more than most people'.[156] Clare Lawson Dick was similarly 'delighted' to be offered £2.15s when she came to the Corporation in 1935.[157] Misses Singer and Lawson Dick were bright, educated young women keen to earn a decent wage and experience independent living in London. For waged women such as these, work at the BBC was a largely positive experience. It was an attractive and prestigious place to work, rates of pay and conditions of service were good, female staff were largely respected and, for those in secretarial and clerical roles with ambition, promotion was a possibility.

Yet although forward-looking and modern in many aspects, so far as waged-women's work was concerned the BBC still conformed to the predominant employment stereotypes of the period and most roles were

[153] R49/31/1, Report on Recruitment of Staff by Ernest Barker and D.B. Mair.
[154] Strachey, *Careers and Openings*, p. 138.
[155] R/49/561/1: Recruitment of Women Clerical Staff, Freeman to Nicolls, 16 March 1934.
[156] Dorothy Torry interview.
[157] Clare Lawson Dick interview.

gender specific. The role of the WSA, with her remit to oversee the female secretarial and clerical staff, echoed many other British interwar workplaces and reveals a continued belief that women workers had separate needs that necessitated a separate line of control. First Miss Banks and then Miss Freeman recruited, inducted and placed weekly paid women staff. They dealt with personal problems, they discussed irritations with line-managers (and the reverse), they established conformity in the way letters were addressed and postage stamps were used. Their domain, the General Office, was the starting place for most new secretarial recruits ensuring their smooth inculcation into the workings of the BBC.

Copy typists, shorthand typists and secretaries were crucial to the BBC with typists performing a fundamental job. The position of personal secretary was also pivotal. In a role that would now be termed a 'PA', these women managed their managers; organising them, caring for them and often advising on and assisting with the work. Clerks also performed the vital role of ensuring that all BBC paperwork was in order. The rapid expansion of the BBC and the constant need for initiative and ideas provided opportunities for bright and ambitious shorthand typists, secretaries and clerks to extend the scope of their employment; many gained salaried status and a small number rose to senior posts. For telephonists and duplicating operators the pinnacle of ambition was a supervisory role. Nevertheless, both were essential services, communications were at the heart of the BBC. By the standards of the day, the BBC's female charwomen, cooks and waitresses were also well treated. However, here too promotion was limited to the confines of the house staff hierarchy.

For a significant number of waged women, work at the BBC was a temporary affair. Clare Lawson Dick recalled how, in the early days, she never expected to stay for long. She simply collected her weekly wage and reserved her energy for going out in the evenings.[158] For many women, a comfortable few years, earning a good wage in a congenial environment was more than satisfactory, promotion to a more challenging job was not something they desired. As *Radio Pictorial* reported, only a few of the hundreds of girls at Broadcasting House had the ambition to become a 'woman who matters' at the BBC, the vast majority were 'content to remain efficient secretaries until they leave to marry'.[159] No records have been kept of the numbers of female secretarial and clerical staff recruited

[158] Clare Lawson Dick interview.
[159] *Radio Pictorial*, 4 January 1935.

by the BBC, neither are there any comprehensive figures on length of service or resignations. However, marriage appears to be the key reason why women left the BBC. In the early years weekly paid women could choose to remain on the staff but, from 1932, they were the main casualties of the BBC marriage bar, as the next chapter will show.

CHAPTER 4

'Only an Exceptional Woman': Married Women at the BBC

On 12 October 1928, as part of the evening series *Questions for Women Voters*, the BBC broadcast a debate 'Should Married Women Work?'[1] The programme, which was produced by Hilda Matheson, pitted Dame Beatrix Lyall against Mrs E.D. Simon. Lyall, an activist in the Mothers Union and a Conservative member of the London County Council put the case against married women's work. Shena Simon, a Liberal member of Manchester City councillor and 'a strong and consistent feminist' maintained that women should be free to choose for themselves.[2] The outcome of the debate is not known, but it was an interesting choice of topic for the Corporation. Unlike many areas of women's work, the BBC in 1928 openly recruited married women and there was no requirement that, on marriage, female staff should resign. However, in 1932 the BBC introduced a marriage bar that was not abolished until 1944.

The BBC's marriage bar was never straightforward. It was at times described by officials as a 'semi-bar'. This was because when it was introduced in October 1932, an allowance was made for those who were 'exceptional' to remain on the staff and many married women kept their jobs. How this distinction was made would cause consistent discomfiture for BBC management, best illustrated by the operation of the Marriage Tribunal (1934–37) where a woman who planned to marry could put

---

[1] *Radio Times*, 12 October 1928.
[2] Earlier that year Simon had successfully led the campaign against Manchester City Council's marriage bar.

© The Editor(s) (if applicable) and The Author(s) 2016
K. Murphy, *Behind the Wireless*,
DOI 10.1057/978-1-137-49173-2_4

forward a case to remain on the staff. The five criteria used to determine her value were loyalty, efficiency, indispensability, career-mindedness and an ability to balance office and married life. Compassionate circumstances were also taken into consideration, underscoring the paternalist nature of the BBC. The minutes of the Tribunal reveal a multitude of viewpoints about married women's lives at this time and their hopes and desires for work. The number of women who came before the Tribunal, though, was comparatively few. During the three and a half years it functioned, 29 women had their cases heard out of whom 13 were dismissed. These small numbers suggest that, in line with the custom of the day, most women who resigned did so voluntarily. The narrative of the BBC's marriage bar, its introduction, application and ultimate abolition, casts new light onto one of the most invidious discriminatory practices of interwar Britain and reveals a range of tensions and contradictions about married women's work.

4.1   THE MARRIAGE BAR IN INTERWAR BRITAIN

Marriage bars were commonplace in the 1920s and 30s. They operated in the Civil Service, in teaching and in banking as well as in large companies such as Cadburys, Great Western Railways, Boots, Sainsbury's and ICI.[3] It had been anticipated that the Sex Disqualification (Removal) Act of 1919 would end the practice. However, when tested in law it became apparent that while the Act might enable married women to work, it did not *entitle* them to do so.[4] Marriage bars were the practical manifestation of the overriding ideology of the period: that married women's sphere was the home. At this time, there was an expectation that once a woman mar-

---

[3] There is a growing historiography on marriage bars, see for instance Alison Oram (1996) *Women Teachers and Feminist Politics 1900–1939* (Manchester: Manchester University Press); Helen Glew (2009) 'Women's Employment in the General Post Office, 1914–1939' (Unpublished doctoral dissertation: University of London); Zimmeck also included information about the marriage bar in her study of women civil servants in the interwar years. Meta Zimmeck (1984) 'Strategies and Stratagems for the Employment of Women in the British Civil Service 1919–1939' *Historical Journal* 27.4, 903–4, 922–3.

[4] Glew, 'Women in the GPO', pp. 125–6. For the impact of the Act on married women see Mari Takayanagi *(2012)* 'Parliament and Women c. 1900–1945' (Unpublished doctoral dissertation: University of London) pp. 55–65. Section One of the Act clearly stated: 'A person shall not be disqualified by sex or marriage from the exercise of any public function, or from being appointed to or holding any civil or judicial office or post, or from entering or assuming or carrying on any civil profession or vocation.' During its passage through Parliament, the Civil Service successfully negotiated exemption, others being the Armed Forces, the Church and the Stock Exchange.

ried she would leave paid employment and become a housewife; her priority being husband and family. It was an ideology reflected in newspapers and women's magazines as well as in the BBC's own broadcasts aimed at its female daytime audience, as Chapter 6 will show.[5] Yet, this created an awkward tension because the interwar years were also a time of expanding employment opportunities for women. Despite the convention to leave work on marriage, many women wanted to retain their jobs. For some, like the educated and trained career woman, this was associated with vocation, ambition and personal fulfilment. For others, the lure of new consumer goods and the quest for the ideal home meant that the benefits of a double-income were tempting, at least until children were born.[6]

The BBC's continued retention of married women, albeit in small numbers, was in keeping with other areas of work. Those in the arts were largely immune from marriage bars which, in part, may explain the BBC's initial progressive attitude. As Ray Strachey noted, 'women musicians, painters, actresses and writers may marry as much as they please, and do in fact marry without abandoning their careers'.[7] Similarly, there was an acceptance of married women working in journalism and advertising. The membership list of the Women's Provisional Club for 1936, for instance, included Mrs Ethel Wood, Director of the advertising agency Samson Clark Co.; Mrs Hilary Blair-Fish, Editor of *Nursing Times* and Mrs Emilie Peacocke, Editor of the Woman's Department at the *Daily Telegraph*.[8]

Married women were not debarred from being MPs. The legal profession, opened to women following the 1919 Sex Disqualification (Removal) Act, also did not operate a marriage bar.[9] Although married women medics employed by local or public authorities might face

---

[5] See, for example, Fiona Hackney (2011) '"They Opened Up a Whole New World": Feminism, Modernity and the Feminine Imagination in Women's Magazines, 1919–1939' (Unpublished doctoral dissertation: University of London) pp. 137–80; Maggie Andrews (2012) *Domesticating the Airwaves: Broadcasting, Domesticity and Femininity* (London: Continuum), pp. 1–81; Adrian Bingham (2004) *Gender, Modernity, and the Popular Press in Inter-War Britain* (Oxford: Clarendon Press) pp. 85–110.

[6] For a discussion on new consumerism and the ideal home see, for example, Deborah S. Ryan (1997) *The Ideal Home through the 20th Century* (London: Hazar) pp. 33–86.

[7] Ray Strachey (1935) *Careers and Openings for Women: A Survey of Women's Employment and a Guide for Those Seeking Work* (London: Faber and Faber) p. 61.

[8] Women's Library: Women's Provisional Club, 5/WPV/3/1, Membership records, September 1936.

[9] Helena Normanton, the UK's first female barrister, successfully campaigned to keep her maiden name. See Judith Bourne (2014) 'Helena Normanton and the Opening of the Bar to Women' (Unpublished doctoral dissertation: University of London) pp. 176–80.

dismissal, married women GPs did not and were often able successfully to juggle work with family life. The LSE took a positive approach towards the employment of married women to the extent that its Director, William Beveridge, introduced a system of child benefit in 1925.[10] Women academics, unlike schoolteachers, were rarely subject to a bar. At least one hundred married women were working in British universities in 1933.[11] A key exception was Liverpool University which, in 1933, introduced a bar, as will later be discussed. Some businesses continued to employ married women. At the John Lewis Department Store the proprietor John Spedan Lewis positively recruited married women managers, particularly in the 1920s, believing them to be more loyal staff and to better understand the needs of customers.[12] In 1939, Ray Strachey in her role as Secretary of the Women's Employment Federation, waxed positively about Vickers Armstrong who were looking to recruit women graduates for their Aeronautical Stress Department. 'Women', Strachey emphasised, 'were employed in all departments ... and there was no marriage bar'.[13]

The examples above all share one similarity, they relate to career-minded women. And this would become key to the philosophy behind the BBC bar. It was observed that there were two classes of married women workers, those who saw their work with the Corporation as permanent and those whose commitment to the BBC was temporary, until they had set up home. This corresponds to what was identified by feminists as the 'meanwhile' attitude of young women in the interwar years, an attitude which was seen to debase all women's work. Strachey in particular lamented the prevalence of this perspective amongst those in the lower ranks of office workers, including clerical staff, shorthand typists and telephonists, where women were eager to swap the routine of their jobs for the idealised notion of the 'nice little home'.[14] Writing in the *Manchester Guardian* in 1928,

[10] William Beveridge (1960) *The London School of Economics and Its Problems 1919–1937* (London: George Allen and Unwin) p. 46.

[11] British Federation of University Women survey 1933.

[12] Judy Faraday (2009) 'A Kind of Superior Hobby: Women Managers in the John Lewis Partnership 1918–1950' (Unpublished MPhil dissertation: University of Wolverhampton) pp. 39–40. A woman buyer for John Lewis contributed a chapter to Margaret Cole's career guidance book where she enthused that you did not have to give up your job on marriage. Margaret Cole, ed. (1936) *The Road to Success: 20 Essays on the Choice of Career for Women* (London: Methuen) p. 242.

[13] Women's Library: Women's Employment Federation: 488 6/WEF/488-4, Advisory Department's Committee Minutes and Correspondence 1938–40, 29 June 1939.

[14] Strachey, Careers and Openings, p. 57.

Vera Brittain bemoaned the belief that 'business' was the chief concern of a man's life but 'marriage' the be-all and end-all of a woman's existence which had been translated into the theory that women's work was only a 'meantime' occupation between school and marriage 'and need be neither carefully studied nor adequately paid'.[15] This resulted in little incentive for training and it also had a detrimental effect on women's pay. Certainly in the GPO, the high turnover of female staff as they left to be married was seen as economically advantageous because it ensured a constant flow of fresh and cheap new recruits.[16] When, in January 1930 nearly 7,000 lower grade female clerks in the Civil Service were asked if they favoured the abolition of the marriage bar, if it meant the end of the marriage gratuity, only 138 women voted in favour. This caused Winifred Holtby to remark:

> Who are the girls who voted for the marriage bar? Nine out of ten swing daily to their offices in suburban trains and trams and buses, carrying in their suitcases a powder-puff and a love-story or 'Home Chat' ... They think that if only they could marry all would be well.

Yet Teresa Davey's study of female Civil Servant clerks in the interwar years shows that they were not the flighty girls Holtby depicted, rather they were bright, hard-working young women who enjoyed their jobs. They did, however, leave on marriage with enthusiasm. Their husbands were often of a higher social class (and had frequently been met through work) and the women looked forward to a companionate partnership as well as sharing a belief that they had earned the right to a 'lovely new home'.[17] At the BBC, while it was the waged female staff rather than the salaried who were more likely to leave on marriage, this was in no way true across the board. Throughout the 1920s and 30s, the complex issue of women and marriage would continue to be a thorny one for the BBC.

---

[15] *Manchester Guardian*, 27 September 1928, quoted in Paul Berry and Alan Bishop eds. (1985) *Testament of a Generation: The Journalism of Vera Brittain and Winifred Holtby* (London: Virago) pp. 125–6.

[16] Samuel Cohn (1985) *The Process of Occupational Sex-Typing: The Feminisation of Clerical Labour in Great Britain* (Philadelphia: Temple University Press) pp. 94–105; Glew, 'Women in the GPO', p. 127. See also Selina Todd (2003) 'Young Women, Employment and the Family in Interwar England' (Unpublished doctoral dissertation: University of Sussex) p. 51.

[17] Teresa Davy 'Shorthand Typists in London 1900–1939' in Leonore Davidoff and Belinda Westover, eds. (1986) *Our Work, Our Lives, Our Words: Women's History and Women's Work* (London: Macmillan Education) pp. 154–9.

4.2   MARRIED WOMEN AT THE BBC: BEFORE THE BAR

In January 1931, the feminist journal *Women's Leader* published an article by Hilda Matheson in which she praised the BBC for its policy on married women. By allowing married women to apply for jobs and by not compelling them to resign on marriage, she maintained, it set an example not always found among public bodies.[18] Matheson suggested that this tolerance and open-mindedness were a result of the BBC being a post-war institution, with a largely post-war staff. And as a young company undergoing fast and haphazard growth, questions about a woman's marital status were not a recruitment priority. Rather, decisions to appoint appear to have been made on aptitude and whether an individual was likely to be an asset to the BBC. It is arguable that the early BBC took a positive decision to employ married women, seeing this as an emblem of modernity.

Although married women were entitled to work at the BBC most chose to resign on marriage. Staff bulletins and magazines frequently celebrated weddings and engagements for example, in May 1928 *The Saveloy* (the original staff newsletter) informed its readers that Isabel Shields, Reith's personal secretary, had left to be married.[19] Of the eight other women whose matrimony was announced (four of whom were engaged to BBC colleagues) only two had elected to stay with the Corporation. One of these was Mary Somerville whose engagement to the journalist R.P. Brown merited a cartoon captioned: 'Happy Mr Peter Brown'. The approaching retirement of Miss Johnson, secretary to the Manchester Station Director, was described as a 'disaster looming ahead' indicating there was often frustration that marriage led to resignation. John Reith, for instance, recorded his sadness that his secretary, Miss Shields, had gone. He noted in his diary her loyal and devoted service over five very strenuous years.[20]

Reith's regret at losing valuable female staff was most pronounced in the case of Olive May, the Telephone Supervisor at Savoy Hill. Her engagement in January 1928 to the Leeds-based engineer Cecil Bottle caused an angry reaction. On hearing the news Reith telephoned his Chief Engineer, Peter Eckersley, and demanded that he take disciplinary action against Bottle for getting betrothed to his 'star operator'. It was left to Lady Reith to telephone Miss May the next day to apologise, explaining

[18] *Women's Leader*, 2 January 1931.
[19] *The Saveloy*, May 1928.
[20] Reith Diaries, 29 February 1928.

that Sir John thought everybody should put the BBC first.[21] May's treachery was forgiven. On the day she left, Reith recorded in his diary, 'Mrs Bottle, senior telephonist since 1923 departing today. I gave her a silver inkstand and went along to see her presents. She has been beyond praise in every way and I regret her going very much.'[22]

In 1932, ahead of the introduction of the marriage bar, BBC management carried out an audit of the approximately 400 female staff. This showed that 31 married women were employed in jobs as varied as multigraph operator, telephonist, registry clerk, shorthand typist and secretary.[23] While some had married since joining the BBC, others had been married when they arrived including two mothers. It is not known if Mary Somerville was included in the original report as selected names were, at some point, redacted but it was Somerville's pregnancy which first prompted the BBC to clarify its position on married women's work. As a senior staff member (she was by now de facto head of School Broadcasting) there was an acceptance that following her wedding in June 1928 she would retain her job. However in November 1928, Somerville announced that she was pregnant. The BBC was now confronted with one of the realities of employing married women, motherhood. Reith made no mention of Somerville's engagement or marriage in his diary but on 22 November he noted, 'Carpendale and wife to tea. We talked about Miss Somerville who is going to have a baby and wants to stay at work.'[24]

Mary Somerville's desire to retain her job forced the BBC to start urgent discussions about its standpoint on maternity.[25] There was no precedent within the Corporation because it was rare for a middle-class woman to be pregnant while at work. Maternity leave was not unknown in the UK; for example, the John Lewis Partnership allowed extended unpaid leave with contributions towards financial hardship being underwritten.[26] However, the widespread existence of marriage bars and the custom to leave the workforce before children were born precluded the possibility of maternity leave in most occupations and professions.

[21] BBC Sound Archive, 87181, 'Reith Remembered', broadcast 21 June 1989.
[22] Reith Diaries, 27 January 1928.
[23] BBC/WAC:R49/371/1: Married Women Policy:1, 1928–35 (hereafter MWP:1), Brief Report on Married Women at Present Working for the Corporation, 25 October 1932.
[24] Reith Diaries, 22 November 1928.
[25] MWP:1, Goldsmith to Reith, 27 November 1928.
[26] Faraday, 'A Kind of Superior Hobby', p. 83.

It was Valentine Goldsmith, in his capacity as Head of Administration, who wrestled with the implications of Somerville's pregnancy. Because the BBC accepted the employment of married women it was not only against 'public policy' but also 'illogical' to rule that motherhood entailed dismissal.[27] Goldsmith recommended that the BBC adopt a scheme, loosely based on the pre-marriage bar arrangements of the London County Council, of four months' full pay and up to a further four months' half pay.[28] There would be no guarantee that the woman could go back to her previous job and if she did not return to work, she would have to repay the money. And while maternity leave might be considered once or twice in a long service, it was reasonable to assume that, 'a woman who is going to have a family of three or more must attend only to it, and give up thought of competing in the wage-earning field on equal terms, and be dependent only on her husband'. In his concluding paragraph, Goldsmith situated the BBC as a progressive institution:

> Looking at the matter as a whole, I feel that any large corporation or commercial organization should take this risk rather than assume a nineteenth century attitude in the present circumstances of women's employment.[29]

This BBC's view of itself as pioneering is further illustrated in a letter from Hilda Matheson to Vita Sackville-West. In May 1929 Somerville, newly on maternity leave, had become seriously ill with tubercular pleurisy and there was concern about how this might affect the unborn child. Matheson, who had just learned of her friend and colleague's illness, wrote, 'It will be sad if all the plans for making her a spectacular vindication of the success of keeping on your job and baby don't come off—poor Maisie.'[30]

Prior to being offered maternity leave, Somerville's case had been referred to the Board of Governors.[31] Here it was agreed that during her 'illness' she was to receive three months' leave on full-pay and up to three months on half-pay. However, while the principle of maternity leave was

---

[27] MWP:1, Goldsmith to Reith, 27 November 1928.
[28] The LCC had introduced a marriage bar in 1923. According to a handwritten note by Goldsmith, prior to the bar, it had operated a system of 8 weeks' full-time/9 weeks' half-time pay. BBC/WAC:L2/195/1: Mary Somerville Staff File, Carpendale to Goldsmith, 7(?) December 1928.
[29] MWP:1, Goldsmith to Reith, 27 November 1928.
[30] Hilda Matheson Letters, 14 May 1929.
[31] BBC/WAC:R/1/1: Board of Governors Minutes, 12 December 1928.

confirmed it was felt undesirable to prescribe fixed regulations for women who became mothers which meant that it was to be discretionary rather than a right. Somerville began her maternity leave in May 1929 returning to the BBC, initially on a part-time basis, in October 1929. The following April she was awarded an above-average pay rise suggesting that her pregnancy and ensuing absence had not adversely affected her standing at work.

Despite the forward-looking treatment of Somerville, the BBC's approach towards married women was beginning to change. For instance, it was the subject of the inaugural meeting of the BBC Debating Society in January 1930. The *BBC Club Bulletin* announced that the society had begun 'triumphantly' with the motion 'That in their opinion women should resign their posts on Marriage'.[32] There is no record of who took part or which side won but it demonstrates that whether married women should retain their jobs was becoming a contested topic.

4.3   MARRIED WOMEN AT THE BBC: A CHANGE IN ATTITUDE

There is no single explanation as to why the BBC's views on married women's work began to change. Rather, there are an assortment of motivations that include economic expediency, social traditionalism, institutional aggrandisement and cynical self-interest. It is also apparent that the driving force behind the introduction of the bar were two senior officials in the Administrative Department, Valentine Goldsmith and his assistant Douglas Clarke. Miss Freeman, newly appointed as WSA also played a leading role.

Economic factors were significant. By 1931, the Depression had taken a deep hold in the UK and, with millions out of work, discussions about married women's employment were heightened. In the early 1930s especially, anger was directed at married women workers who were seen to be taking jobs away from single women as well as from unemployed men. Even the socialist and feminist Margaret Cole acknowledged the 'very real fear' among men if married women were allowed or encouraged to go on earning or holding down jobs, a fear shared by their wives.[33] It was in this light that married women's work was discussed three times at the BBC's Control Board in 1931. The all-male executive tussled with questions such

---

[32] BBC Club Bulletin, February 1930.
[33] Margaret Cole (1938) *Marriage: Past and Present* (London: Dent and Sons) p. 202.

as whether women whose husbands were in work should be refused BBC posts and how married women already on the staff should be treated.[34] To get a better understanding of the issues it was decided that Clarke, Goldsmith's assistant in the Administrative department, would investigate practices in other firms.[35] Clarke was diligent and in September 1932 he reported to Goldsmith that he had held long interviews with the London Life Association, the Ministry of Labour, Imperial Chemical Industries, the Prudential Assurance Company, the National Provincial Bank, the London County Council, Marconi and the Underground Railways of London.[36] He had also sought corroboration on the issue from the Civil Service. With the exception of the Underground Railways of London, Clarke affirmed, all the companies terminated their women staff on marriage, giving as their reasons 'principally the labour market', although most also expressed the personal view that married women 'could not well carry on a business and run a home'.

This view of the double-burden was widely held. It was not deemed practical for a respectable woman to attempt both to work and run a home. In August 1933, for instance, when the press got a whiff of the BBC's marriage bar, the *Glasgow Evening Citizen* bellowed:

How can a woman possibly do justice to her home and to her family if she has to devote her attention to another job? Equally, how can she fulfil her obligation to her employer if her mind is preoccupied, as it must be, with domestic affairs? No more than a man can a woman serve two masters.[37]

Domesticity was used to enforce the marriage bar legally. For example, in 1925 a challenge to the bar was brought by a female teacher in Dorset. When the case of *Short v Poole Corporation* went to appeal, Poole Corporation, which won, maintained that 'the duty of married women was primarily to look after her domestic concerns, and they regarded it as impossible for her to do so and to act effectively and satisfactorily as a teacher at the same time'.[38]

---

[34] BBC/WAC:R3/3/7: Control Board Minutes, 5 February, 22 September, 15 December 1931.

[35] Control Board Minutes, 15 December 1931.

[36] MWP:1, Clarke to Goldsmith, 23 September 1932.

[37] *Glasgow Evening Citizen*, 31 August 1933.

[38] As quoted by Erna Reiss 'Changes in the Law' in Ray Strachey ed. (1936) *Our Freedom and its Results by Five Women* (London: Hogarth Press) p. 99.

Clarke's report gives an intriguing insight into fluctuating attitudes towards married women's work which encompassed social, economic and cultural concerns. On the 'Pro' side he agreed that women 'presumably have the right to live as they think fit' and were therefore entitled to seek employment or remain in employment after marriage. He also concurred that married women might obtain an experience and balance 'lacking in certain single women' and that they might be more stable members of staff, as they would not have 'the restless outlook of so many girls who are contemplating marriage'. On the 'Contra' side he surmised, with the labour market in its present condition, it was unfair for married women who were supported by their husbands to compete against unmarried women who needed to earn a living. Clarke also vented the view that it would seem impossible for a married woman to work and at the same time maintain a reasonable home for her husband and her family, 'either her work in business must suffer, or her health, or her husband, or her children'. He pointed out that, through remaining at work, she might not have children, which could be bad for the community and herself. Clarke ended his report by determining that the arguments against retention outweighed those for retention though he concurred that it was 'of course a most difficult matter to decide upon'.[39] Whilst the main thrust of Clarke's report was economic and social there was also a hint at what would become an overriding rationale for the introduction of the BBC marriage bar—a perceived difference in attitude amongst women staff towards their work. A series of emotive memos between Clarke and Goldsmith reveal this new reasoning, sparked by a reconsideration of the BBC's system of Marriage Leave.

Like the Civil Service, the BBC offered a gratuity to women who left on marriage, in lieu of pension.[40] Anyone who had paid into the pension scheme was entitled to receive a refund of their contributions plus interest. In addition, the BBC operated a system of Wedding Present and Wedding Leave which was granted equally to both male and female staff. The Wedding Present was a gift of between £5 and £10 for weekly paid staff (depending on length of service) and one-thirtieth of annual salary for those who were monthly paid.[41] Wedding Leave was an extra week's holiday. In August 1932, Clarke discovered this was being abused with

---

[39] MWP:1, Clarke to Goldsmith, 23 September 1932.
[40] The Civil Service offered a better deal, as would become apparent.
[41] MWP:1, Goldsmith to Salaries Clerk, 25 August 1927.

the honeymoon allowance being taken even though there was an intention to resign from the BBC soon afterwards. Angered by what he saw as
an attempt to play the system these 'girls', he declared, wished to remain
with the BBC only for a short period, in order to add to their means,
and so were, 'making a convenience of the Corporation and in certain
cases causing inconvenience to us'.[42] Clarke informed Goldsmith that,
with Freeman's agreement, the extra week's leave should no longer be
given to such women. Frustration was focused upon one woman, Miss
Robertshaw, whose work with the BBC had not been in 'any way note-
worthy' and who he and Freeman would be rather glad to replace.

This raised two apposite issues: women's intention to stay at the
BBC after marriage and their aptitude. It begged the question: should
those considered to be inadequate in their work have the right to stay?
Goldsmith's response is telling. He both supported Clarke and Freeman
in their desire to tighten up Marriage Leave and elaborated on Clarke's
distinction between women's motivation for work:

> The first [class of woman are] those who intend to marry and remain in the
> ranks of women workers permanently ... i.e. they regard themselves equally
> with their husbands, as workers, and not as domestic partners in the mar-
> riage ... The second class consists of those who have no intention of being
> women workers save for their financial needs during a temporary period of
> getting a home together, whose outlook is different and whose mind is not
> here but in their homes.[43]

He then posed the conundrum whether 'without making an arbitrary
rule' the BBC could 'put an end to the short-time convenience worker,
who is less interested in her job here' though he did agree that any decision would be hard and 'must depend on past work and attitude'. As
he reasoned, Freeman would have refused Robertshaw's continuation of
work after marriage had she been free to do so but, 'our marriage rules
bereft her of this freedom'.

That the introduction of the BBC's marriage bar was ignited by this
particular incident is signalled by Freeman who recalled that it was 'the
case that made me first query the Corporation's policy with regard to
the automatic retention of women staff after marriage'.[44] Indeed, when

[42] MWP:1, Clarke to Goldsmith, 16 August 1932.
[43] MWP:1, Goldsmith to Carpendale? 26 August 1932.
[44] MWP:1, Freeman, undated and unsigned memo.

the introduction of the bar was promulgated to women staff it was made clear that 'certain cases' had led the BBC to reconsider its position.[45] So, the reappraisal of Marriage Leave appears to have been a catalyst for the BBC marriage bar, a bar which would enable the retention of desirable women while allowing the Corporation to dispense with certain women it did not want. Freeman was unambiguous that this was a factor:

> ... sometimes the girl in question was not particularly efficient and we were glad of the chance to get rid of her. In fact by not allowing automatic retention after marriage we were enabled to dispense with some of the less satisfactory employees whose work was not so poor as to justify dismissal.[46]

This self-interest on the part of the BBC makes uncomfortable reading although there are no details of specific cases of women removed in this way.

Alongside self-interest and economic and societal factors, there appears to have been a further trigger for the BBC marriage bar: the Corporation's growing sense of itself as an esteemed British institution.[47] By 1933, Hilda Matheson was confident that it was now 'a part of national and international machinery'.[48] Prior to Clarke's 1932 report the BBC had not fully considered its policy on married women's employment in the light of other workplaces. Now that marriage bars were acknowledged as part of the prevailing cultural orthodoxy, a BBC marriage bar might accordingly add to the Corporation's sense of conformity and respectability. It is probably no coincidence that the introduction of the bar coincided with the move from the ramshackle offices of Savoy Hill to the grandeur of Broadcasting House in 1932.

Yet, aligning itself with the establishment created a dilemma for the BBC: how might its reputation as a progressive organisation be affected by the introduction of a bar? According to Goldsmith, Reith wanted to discourage married women workers but had hesitated to change the rule because, 'in women's papers our outlook has been upheld as a good one facing modern facts, and any change would have immediate outside notice and be widely commented upon'.[49] Goldsmith was right to be concerned.

[45] MWP:1, BBC Marriage Bar Statement, 15 August 1933.
[46] MWP:1, Freeman, undated and unsigned memo.
[47] D.L. LeMahieu (1988) *A Culture for Democracy: Mass Communications and the Cultivated Mind in Britain between the Wars* (Oxford: Oxford University Press) p. 189.
[48] Hilda Matheson (1933) *Broadcasting* (London: Thornton Butterworth) p. 207.
[49] MWP:1, Goldsmith to Carpendale? 26 August 1932.

4.4   MARRIED WOMEN AT THE BBC:
THE INTRODUCTION OF THE MARRIAGE BAR

A few days before the Board of Governors met to approve the introduction of the marriage bar there was worrying news for the Corporation. Feminist campaigners appeared to have learnt of the BBC's plans for change. On 21 October 1932, Mary Somerville informed Clarke that the 'Suffragette element' of women's organisations were seeking a test case on the grounds that it was illegal to force a woman to resign on marriage. Somerville thought it likely that, should the Corporation come to a decision that women should either be dismissed or required to resign, the test case would be the BBC. This 'even if abortive', Clarke stressed, 'might have very unpleasant results to us'.[50] Since the early 1920s, feminist groups such as the Open Door Council, the Six Point Group, the National Union of Societies for Equal Citizenship and the Women's Freedom League (WFL) had actively campaigned against the marriage bar. It seems probable that Somerville had that day attended a conference at Caxton Hall organ-ised by the WFL on 'The Position of Married Women', the precursor of the Campaign for the Right of the Married Woman to Earn.[51] Although the subject of the BBC was raised at the conference, in the event, it was Liverpool University that would become the focus, as we shall see.[52]

Despite jitteriness, the BBC marriage bar was approved by Governors on 27 October 1932.[53] As yet, though, there was no official announce-ment and it would take eight months of painstaking deliberations to finesse the wording of the 'Statement to Women Staff'. By the time this was delivered there had been a major restructuring within the BBC so it was not Valentine Goldsmith but Basil Nicolls, as the new Director of Internal Administration (DIA), who introduced the ruling. On 15 August 1933 the BBC's female employees were informed that, while there was to be no definite bar, in future the retention of married women was to be regarded as exceptional and dependent upon the circumstances of individual cases. In coming to this decision the Corporation had:

> ... largely been guided by a belief that only an exceptional woman, with adequate material resources, can perform her duties satisfactorily as a

[50] MWP:1, Clarke to Goldsmith, 21 October 1932.
[51] Mary Somerville Private Papers, Speech to teachers 1938.
[52] Women's Library/Women's Freedom League:2/WFL01–15, National Executive Committee Meeting, 22 October 1932.
[53] BBC/WAC:R/1/1: Board of Governors Minutes, 27 October 1932.

whole-time servant of the Corporation, while attempting to fulfil the cares and responsibilities of a young family.[54]

The BBC had hoped to avoid scrutiny through notifying staff by memorandum rather than by incorporating the new ruling into the Staff Agreement or the Standing Instructions.[55] Unfortunately for management, somehow the 'Statement to Women Staff' got into the hands of the press. One of the most intriguing features of the ensuing coverage is how differently— and how wrongly—it was interpreted. The *Evening News* claimed that the 'Resign-on-Marriage rule' had been relaxed.[56] The radio columnist Collie Knox, writing in the *Daily Mail* declared that, 'BBC Girls May Marry'.[57] Only the *Daily Mirror* got the story broadly correct, covering the issue with both a full-page article and an editorial.[58] Under the headline 'BBC Dictatorship: Married Women's Rights', the barrister and political activist Helena Normanton railed as to whether, 'some policy of more or less compulsory celibacy' was on its way. 'Fair play is such a jewel', she protested, 'that it would make us all very uneasy to feel that there is any possibility of one rule ... for the highly-placed woman, and another and harsher for the stenographer or translator.'[59]

Considering Helena Normanton's sharp criticism of the BBC it is surprising that these details failed to filter into the Campaign for the Right of the Married Woman to Earn which was then at the peak of its activity and with which Normanton was involved.[60] Liverpool University, which had introduced a marriage bar in February 1933, was its prime target.[61] The newly married Dr Margaret Miller, an eminent academic at the university, was the

[54] MWP:1, Statement to Women Staff, 15 August 1933.

[55] MWP:1 Clarke to Goldsmith, 21 October 1932.

[56] *Evening News*, 28 August 1933.

[57] *Daily Mail*, 29 August 1933.

[58] Reith was horrified by the coverage, noting in his diary that 'the "Daily Mirror" rag had a great stunt about our attitude to married women. A gross misrepresentation of the facts.' Reith Diaries, 28 August 1933.

[59] *Daily Mirror*, 28 August 1933.

[60] University of Liverpool: Margaret Miller Papers: D384/2/83-112. See, for example, letters dated 13 and 14 June between Normanton and Florence McFarlane of the Six Point Group.

[61] For an analysis of Liverpool University's marriage bar and its links with the Campaign for the Married Women to Earn see Jennifer Bhatt (1995) 'Margaret Miller and the Campaign for the Right of the Married Woman to Earn' (Unpublished MPhil dissertation: University of Leicester).

main casualty of the bar.[62] A political activist, Miller refused to go quietly and her case led to the formation of the Campaign. On 14 November 1933, a Mass Meeting of the Campaign for the Right of the Married Woman to Earn was held at Central Hall, Westminster. It attracted 3000 women and was addressed by, amongst others, Nancy Astor MP and the writer Rebecca West.[63] In her speech, Mrs Pethick-Lawrence, President of the WFL, did mention the BBC, but in connection with its appointment and quick dismissal of a woman announcer.[64] The Mass Meeting received blanket coverage in the press and prompted Liverpool University to reconsider its position, abolishing its marriage bar in March 1934.[65] It is interesting to speculate how the BBC might have reacted if it had been made a focus of the Campaign. In the event, it dodged the spotlight and so was not put into the awkward position of defending itself in the face of intense feminist scrutiny.

By the autumn of 1933, the BBC marriage bar was securely in place, with female employees informed of its parameters. The tricky problem BBC management now faced was how to decide which women would be exempt.

4.5   THE BBC MARRIAGE TRIBUNAL

On 12 May 1933, Reith met to discuss the issue of the retention of married women staff with Mary Somerville, Miss Freeman and the BBC governor, Mary Agnes Hamilton, an occasion of sufficient import to merit an entry in his diary.[66] In her record of the meeting Somerville noted that the notion of the 'exceptional' woman had been discussed as had the small tribunal which would consider the case of anyone 'desiring to know whether she would be regarded as an exceptional case'.[67] It seems

---

[62] There is a suggestion that Miller was viewed as difficult and challenging to the male hierarchy, her dismissal in part due to a strained relationship that had developed between herself and two senior male colleagues. See, for example, Margaret Miller Papers, Margaret Miller to Mrs Adami, 11 December 1932.

[63] Groups that attended included the Association of Assistant Mistresses, the Association of Women Clerks and Secretaries, the National Union of Women Teachers and the National Association of Women Civil Servants. Other affiliated groups included the National Women's Citizen's Association; the Fabian Society Women's Group, the Soroptimist Club; the Association for Moral and Social Hygiene and the YWCA.

[64] *Daily Telegraph*, 15 November 1933.

[65] Margaret Miller Papers: The University of Liverpool: Memorandum to Special Purposes Committee, Employment of Married Women, 14 March 1934.

[66] Reith Diaries, 12 May 1933.

[67] MWP:1, Somerville to Nicolls c May 1933.

likely that the BBC's Marriage Tribunal was inspired in part by a procedure that existed within the Civil Service. In 1931, the Report of the Royal Commission on the Civil Service (the Tomlin Commission) had been published, one of the areas considered was the retirement of female civil servants on marriage. While it was agreed the marriage bar should be retained, the Commission identified a difference in attitude towards the higher and lower grades of Civil Service women indicating that 'a considerable body of opinion' was in favour of treating the higher grades differently.[68] It was therefore agreed that an exceptional woman could be retained on marriage if it was deemed in the public interest, individual cases going before the Treasury.[69] Mary Agnes Hamilton had been a member of the Royal Commission and it was Hamilton who mooted the idea of the BBC Tribunal.[70]

Discussions about what made a woman 'exceptional' continued at the BBC for many months with 'indispensability' the prime focus. According to Nicolls, the DIA, this referred to the applicant's special experience and/or the difficulty of replacing her.[71] He used Mary Somerville as a prime illustration. Although long married, if her case were to come up afresh, he surmised, it would almost certainly be held that her special experience in school broadcasting was of great value and made her very difficult to replace. Nicolls also pointed out that since the members of the staff with the most valuable experience were likely to be the most highly paid it was inevitable that the exceptions to the rule should mainly occur among the senior staff. This would be evident in the rulings of the Marriage Tribunal; no salaried woman was ever required to resign.

The BBC Marriage Tribunal heard its first case on 28 December 1933. Five criteria had eventually been decided upon and a points system introduced. These were:

1. Special value of experience, making replacement difficult or undesirable. (Maximum 100 points)
2. Compassionate circumstances. (Maximum 50 points)

---

[68] Glew, 'Women in the GPO', p. 127.

[69] Between 1934 and 1938, eight women in the Administrative Grades were retained on marriage. Hilda Martindale (1938) *Women Servants of the State, 1870–1938* (London: Allen and Unwin) p. 156.

[70] Mary Agnes Hamilton (1944) *Remembering my Good Friends* (London: Jonathan Cape) p. 264.

[71] MWP:1, Nicolls, c January 1934.

3. Long service and general efficiency. (Maximum 50 points)
4. Character as bearing on the strain of combining married life with office work. Maximum 50 points)
5. Intention of making a career in the BBC. (Maximum 50 points)[72]

Unless there were conflicts of interest, the Tribunal was made up of the DIA (Nicolls); the Establishment Officer (Clarke); the WSA (Freeman) and two independent assessors, one male, one female. These were senior staff with at least five years' service and from a different division to that of the woman whose case was being heard. Aware that the independent assessors might have a personal stance on the marriage bar, at the beginning of each meeting the Chairman warned members that the Tribunal had been summoned to interpret a definite policy, so that 'a woman member who was entirely opposed to the policy might and should rightly vote for not retaining an applicant'. The woman whose case was before the Tribunal was not herself present, rather documents were submitted from her managers and from Freeman, who spoke to each individual beforehand to garner the particular details of her situation. Representation by a friend or colleague was agreed in principle in March 1937, shortly before the 'experiment' of the Tribunal came to an end.[73]

The Marriage Tribunal documentation makes engrossing reading, both the minutes of the hearings themselves and the behind-the-scenes discussions. The scrutiny given to the minutiae of the women's lives is unsettling, it is hard to imagine the personal life of a male employee being probed in the same way. Looking through the 29 cases, three of which came before the tribunal twice, one is instantly struck by the arbitrary nature of the scoring system which was largely dependent on the report of the manager and the whim of the board. Indispensability was the top criterion for retention so it is something of a surprise that Lilian Lord, a salaried clerk in the Supplementary Publications Department scored most highly for Criterion One, 'Value' (97 out of 100) far above that of Jean Melville the acclaimed accompanist who scored only 70.[74] The BBC's self-interest is palpable. The women were assessed by their usefulness to the

[72] BBC/WAC:R49/372: Married Women Policy: Tribunals, 1934–37 (hereafter MWPT), Tribunal Minutes, 30 January 1934 and all subsequent tribunals.
[73] MWPT, Pym to Freeman, 23 March 1937.
[74] Tribunals, 16 January 1935, 17 August 1934.

Corporation and the difficulty of finding a replacement. This was made even clearer in December 1934 when greater leniency was introduced:

> Hitherto indispensability has been interpreted as meaning that the individual had some special experience or qualities which it would be extremely difficult for the BBC to obtain by replacing her, and the mere inconvenience of replacement was not taken into consideration. In future, more weight may be given to such inconvenience.[75]

Subjectivity was even more of an issue when it came to Criterion Four 'Character'. For this, the panel were in the hands of Freeman who directed them with comments such as, 'WSA said that she considered that Miss A was a level-headed sort of girl and not the kind to give way under the strain' and 'WSA pointed out that Miss B was a girl of the class that regarded it as natural to have to work for a living, and that there was no doubt that she would have no difficulty in running her home in addition to her work'.[76] While most women who came before the tribunal were considered capable of combining home life and office work, a handful were not. For instance Freeman was clear that: 'Miss C's health record is not good and WSA expressed doubt as to whether she would find it easy to combine married life with her work here'.[77]

These considerations emphasise the expectation in the 1930s that women were the home-makers. If a married woman worked, it was she who had to negotiate the double-burden of two jobs. The reason why the vast majority of BBC women were viewed as capable of doing both was either their fortitude or their ability to pay for domestic help. Many ran their own flats, for example the drama producer Barbara Burnham already employed a housekeeper. For others, it was felt that the fact of being married would lessen the domestic burden. So it was noted that one woman, a Studio Executive Clerk, 'has for some years had domestic responsibility in her own home and it appears that that is likely to be lessened rather than increased when she sets up house with her husband'. For a secretary in the Office Administration Department 'marriage would ease the situation by enabling her to employ a servant'.[78] For yet another, the burden would be eased because she would no longer have sole care of her widowed

---

[75] MWP:1, Note on the Marriage of Women Staff, Nicolls, 21 December 1934.
[76] Tribunals, 16 January 1935, 11 May 1936.
[77] Tribunal, 5 March 1937.
[78] Tribunals, 25 February 1937, 5 August 1936.

mother.[79] These cases show that for the majority of BBC women, it was considered possible to both work and run the marital home provided there was adequate domestic support. In none of the BBC documentation was there any suggestion that husbands might share the chores.

'Long Service' and 'Career', Criteria Three and Five, were the most straightforward to address. Of the ten women who applied to the Tribunal with less than four-and-a-half years' service only one was retained, Mary Allan, who had been appointed specifically to run the Television Make-up and Wardrobe Department. Similarly, all those who had been awarded their ten-year bonus, whether salaried or waged, received unanimous votes for retention, indicating that loyalty was an important consideration for the BBC. Fifteen women voiced their intention of remaining with the Corporation even if their husbands had sufficient finances to support them, only two of whom were subsequently required to resign, showing that commitment to a career was viewed positively. *Radio Pictorial* applauded the 'intelligence and humanity' of the BBC's matrimonial rules, which were 'never more plainly vindicated' than in the case of Barbara Burnham:

> It is one thing to engage a married woman whose husband can support her while single girls as well qualified are seeking the same job. It is another to dismiss a woman whose temperament and long training fit her for the almost unique position which she occupies, just because she wants to marry.[80]

The criterion that caused the most soul-searching for the Tribunal was 'Compassionate Circumstances' which related both to caring responsibilities and to the means of the prospective husband. Nearly all the women provided evidence that their fiancés were not earning enough to 'keep two people in any degree of comfort'.[81] One was a ballet dancer whose income was subject to perpetual fluctuations; another, an osteopath in a new practice; a third was an 'architect in the London County Council (LCC) with a salary of approximately £3 a week, with prospects of a slow increase'.[82] Others included a carpenter, a transport foreman and a stoker in the Navy. For many women, and their fiancés, there were financial dependents: an elderly father, widowed mothers, younger siblings who were students or in precarious jobs. One applicant was suffering from a disease 'which did not at present affect her

[79] Tribunal, 19 January 1937.
[80] *Radio Pictorial*, 10 June 1938.
[81] Tribunal, 17 July 1937. Some Tribunals heard more than one case.
[82] Tribunals, 17 July 1937, 5 May 1935, 11 May 1936.

efficiency, but was such that her doctor had strongly urged her to get married as the best method of effecting a cure'.[83] The BBC's attention to these considerations reveals its paternalism and reiterates its reluctance to introduce a blanket bar.

One of the most vociferous arguments against marriage bars in the interwar years was the enforced singlehood they imposed.[84] For feminists the issue was human rights, for others it was seen as wrong to deny those who wanted a career the pinnacle of womanhood: marriage and family life. March 1935 saw the BBC wrestling with the issue. Two women who had failed the Marriage Tribunal the previous year had reapplied, the financial circumstances of their prospective husbands having worsened. In his weighing up of their second appearance Nicolls was clear that under the ruling, they should not be kept on. Yet he acknowledged that, 'Here we are up against the very difficult policy question of our action preventing early marriages'. For one of the two women, the choice was to marry and be permanently extremely hard up, to stay on unmarried, or to take the risk and leave, hoping to find suitable work elsewhere. The second woman's situation was more serious in the eyes of the BBC. She was already 35 so her chance of marrying was 'a relatively poor one' but by staying on, she was running a risk of never getting married. The compassionate circumstances were such that she could not get married unless she was able to earn her own living and support certain dependants. For Nicolls the central question was whether the BBC was 'better served in the long run by Miss X or Miss Y as an embittered, because compulsory, spinster, or by her as a contented married woman allowed to remain on the staff?'[85] It was agreed that both women could stay.

Another worry raised by forcing women to postpone matrimony was the possibility of a secret marriage or, worse still, compelling a couple to 'live in sin'. Speaking at the Mass Meeting for the Right of the Married Woman to Earn in 1933, Nancy Astor declared that, because of marriage bars 'thousands of women nowadays are secretly married, or, worse still, living with the men they ought to be married to'.[86] It is known that there were cases of teachers and assembly-line workers who hid their wedding

[83] Tribunal, 17 July 1934. No details were given about the disease.
[84] See, for example, Katherine Holden (2007) *The Shadow of Marriage. Singleness in England, 1914–1960* (Manchester: Manchester University Press) pp. 15–16.
[85] MWPT, Nicolls to Carpendale, 6 March 1935.
[86] *Daily Mail*, 15 November 1933.

rings to keep their marriages secret.[87] The LCC informed the BBC that they had occasional cases of women who were found to have got married secretly and who were then sacked.[88] The London Passenger Transport Board (who by 1937 had instituted a bar) also recognised that many unreported marriages took place.[89] The BBC itself was not immune to the problem. Freeman herself acknowledged that, 'at different times, three women have been dismissed when it was found that they were married'.[90]

While Reith maintained a distant interest in the Marriage Tribunal, the BBC Governor Mary Agnes Hamilton played an active role. It had, after all, initially been her idea and she was sent the minutes of all Tribunals 'so that she might have an opportunity of urging more lenient treatment, or reconsideration'.[91] Why Hamilton, with her feminist leanings, supported the BBC's marriage bar is puzzling. Prior to her appointment as a Governor, she had served as Labour MP for Blackburn 1929–31. The official Labour Party line was against marriage bars, although a significant minority of members were in favour of them.[92] Perhaps her support for the BBC bar was a result of her passionate trade unionism. In Parliament she had frequently attacked her party's failure to solve the unemployment crisis. Her ambivalence about married women's work is apparent in her book *Women at Work* in which she represented the two polarised views on the subject.[93] And it was Hamilton who first suggested that the BBC Marriage Tribunal should be replaced by a simpler Civil Service-style system whereby only women above a certain grade or salary could be considered for retention, a system that was ultimately adopted.[94]

It had gradually become apparent that the Marriage Tribunal was untenable. On 16 July 1937 it heard its final case and was then suspended pending discussions on its future. Not only was it viewed as work-intensive for BBC management but it put the women whose cases were heard under

[87] Oram, *Women Teachers*, p. 56; Miriam Glucksmann (1990) *Women Assemble: Women Workers and the New Industries in Inter-War Britain* (London: Routledge) p. 223.

[88] BBC/WAC:R49/371/2: Married Women Policy:2, 1936–July 1939 (hereafter MWP:2), Message from W.H. Young, LCC, 29 July 1937.

[89] MWP:2, Clarke to Freeman, 14 July 1937.

[90] MWP:2, Freeman to Pym, 7 October 1938.

[91] MWP:2, Reith to Nicolls, 26 September 1935.

[92] Pamela Graves (1994) *Labour Women: Women in British Working Class Politics 1918– 1939* (Cambridge: Cambridge University Press) pp. 126–31.

[93] Mary Agnes Hamilton (1941) *Women at Work: A Brief Introduction to Trade Unionism for Women* (London: Routledge) pp. 5–6, 164–8.

[94] MWP:2, Carpendale to Pym, 30 June 1937.

immense strain. It had also created an awkward issue in connection with severance pay because 'termination' was initially at a higher rate than 'resignation', something it took many months to resolve.[95] In addition, it was seen as relatively ineffectual; in the three and a half years it operated only 13 women had their services terminated whereas 79 women had elected to resign on marriage during this time.[96] It is impossible to know how these numbers relate to resignations prior to the introduction of the marriage bar because no statistics were kept.

What is certain is that 1933 marked a climate change for married women staff, the antipathy towards the 'non exceptional' would have encouraged many to leave. Freeman's fact-finding interviews with married women in 1932 definitely had the effect of prompting a number of departures, the scrutiny of their situations causing discomfort.[97] A query from Reith in March 1933 about whether numbers had been 'reduced' garnered the response from Freeman that five had left with two about to leave.[98] Freeman was non-committal as to whether the bar had increased resignations amongst unmarried staff, stressing that no woman had indicated this to her although she did advise those with weak cases against going before the Tribunal.[99] Nicolls was less circumspect. Now Controller (Administration) he claimed:

> ...undoubtedly [the policy had] been the cause of many girls deciding to give up work on marriage, who would probably have stayed on, without it being financially necessary for them to do so: in fact the most remarkable point has been the fewness of the applications.[100]

4.6   Married Women at the BBC Post-1932

For married women who remained at the BBC, whether at Head Office or in the Regions, life continued as normal with the same prospects for increments and promotions as other female staff. This applied both to those

[95] See, for example, MWR:2, Nicolls to Pym, 21 January 1937.

[96] MWP:2, Staff Records to Clarke, 16 July 1937. This represented a yearly average of 4.5 per cent of the total female staff.

[97] MWP:1, Brief Report on Married Women at Present Working for the Corporation, 25 October 1932.

[98] MWP:1, Married Women, Draft Letter, early March 1933.

[99] MWP:2, Freeman to Pym, 7 October 1938.

[100] MWP:2, Note on Policy in regard to Married Women, prepared by Nicolls for Lady Bridgman, June 1936.

who passed the Marriage Tribunal and to those who had married before the bar was introduced. For instance, Alice Wright, who was successful at a Tribunal in 1935, became Deputy Music Librarian in 1937. Most continued to use their maiden names. Of the 38 married women listed as working at the BBC in February 1937, 26 were 'Miss', so there was little indication that their marital status had changed.[101] Nearly all those referred to as 'Mrs' had married prior to the introduction of the bar.

A small number of married women subsequently resigned, very possibly to start a family. By October 1938 five of the 16 women who had passed the Tribunal had left; in addition two women had requested maternity leave.[102] These were Mrs Benham, a clerk in Programme Finances and Mary Adams, a senior talks producer, whose maternity is discussed in Chapter 6.[103] Another married woman and mother on the staff was Caroline Towler. One of the supposedly clear principles of the BBC marriage bar was that married women would no longer be recruited by the Corporation. Despite this, in the summer of 1933 Reith personally sanctioned the return of Mrs Towler to the BBC. As Miss Banks she had been Miss Freeman's predecessor as WSA and was a great favourite of Reith's. Since her resignation in 1931 her naval officer husband had become unemployed and she appealed to her old boss to be allowed back.[104] Clearly, when it was decreed by the Director General, a married woman could be employed. Towler, now a mother to two young children, including a four-month-old baby, was found a salaried role as Night Hostess, a glamorous position that involved meeting and greeting the evening's broadcasters. *The Evening News* assured its readers that her children would not be missing her much 'for her duties do not start until 6.30 pm. They finish at 11 pm.'[105] The press were evidently intrigued by working mothers at the BBC, the children of Mary Somerville and Sheila Borrett were also deemed newsworthy as later chapters will show. This was emblematic of the often contradictory coverage in the press, on the one hand berating married women who worked and on the other marvelling at their modernity.

The latitude that was shown to Caroline Towler was extended to the BBC's temporary female staff. Many occupations and industries which

---

[101] MWP:2, February 1937.
[102] MWP:2, Freeman to Pym, 7 October 1938.
[103] There are no details of Mrs Benham's maternity leave.
[104] Reith Diaries, 9 June 1933, 'Miss Banks to see me, in a very bad way'.
[105] *Evening News*, 26 August 1933.

imposed marriage bars made exceptions for casual labour such as holiday cover or seasonal work. Married teachers, for instance, were often accepted for supply work and the Post Office also employed married women as temporary staff during busy times.[106] The BBC followed this convention and used temporary agency staff to cover sick leave, holiday leave and periods between appointments, nearly always married women.[107] This was because it was 'almost impossible to get unmarried girls to do temporary work as, owing to the shortage of the market at the present time, they are snapped up immediately for permanencies'. Many of the women had previously worked at the BBC and were seen as particularly valuable because they not only understood the Corporation but were known to be reliable and trustworthy. However, BBC supervisors increasingly found themselves in conflict with the Administration Division over their use of married women staff in a temporary capacity, especially when contracts were for lengthy periods of time. The situation was clarified in June 1937 with new guidelines that married women could be employed but only for a maximum of six months at a time and for no more than eight months in any given year.[108]

By mid-1937, it was apparent that the regulations surrounding the marriage bar were not working. But how might the BBC progress? What, if anything, should replace the Tribunal? It was Miss Freeman who had the closest dealings with women staff and it was Freeman who, in 1937, voiced the first concerns about the continuation of the bar.

4.7   THE ABOLITION OF THE BBC MARRIAGE BAR

Miss Freeman had initially supported both the marriage bar and the Marriage Tribunal, indeed she had been pivotal to their introduction, but by March 1937 she had changed her mind. She informed William St John Pym (who, in 1936, had replaced Nicolls as Director of Staff Administration) that she would now 'welcome an experiment on the other side, namely the definite lifting of the bar'.[109] Freeman gave two reasons for her modified opinion: the shortage of good secretarial workers and because it was 'the only subject on which there is a justifiable feeling

---

[106] Peak Frean, for example, which enforced a strict marriage bar for permanent employees, welcomed married women during peak times, Glucksmann, *Women Assemble*, pp. 107–8; Oram, *Women Teachers*, pp. 69–70; Glew, 'Women in the GPO', pp. 161–3.

[107] MWP:2, Clarke to Freeman, 8 February 1937.

[108] MWP:2, Extract from Mrs Winship's file, 4 June 1937.

[109] MWP:2, Freeman to Pym, 9 March 1937.

of discontent among the women staff'.[110] Aside from Freeman's comment, there is scant information from BBC women themselves about their attitude towards the bar. A meeting organised by the Association of Women Clerks and Secretaries in December 1937, for example, was very poorly attended even though the flyer (which was sent to all women staff) included the marriage bar.[111] A key reason why there was no agitation against the BBC appears to be because it did not affect salaried staff or 'career women'. In teaching and the Civil Service, for example, it was high-achieving women who were hit the hardest and it was they who were the most vocal opponents.[112] The case of Margaret Millar exemplified this. It was because she was a highly articulate professional woman that her dismissal became a cause célèbre.

At the BBC, there was a single reference to the marriage bar in *Ariel*. The October 1937 edition included a short editorial on the marriage bar which indicated that the Tribunal was to be abolished and which approved the Civil Service practice that was to replace it. Joyce Morgan, secretary to the Editor of *The Listener*, was 'Guest Editress' of the issue. Encouraged to add her viewpoint to any article that caught her eye, she denounced the new proposals as hitting 'the very people whom one would suppose any "marriage scheme" would aim at assisting'; women who, for economic reasons, needed to go on working. Morgan described the Civil Service practice under consideration as the 'marriage discouragement scheme'.[113]

The future of the marriage bar and the possibility of a Civil Service-style scheme were due to be discussed by the Control Board and the Board of Governors in early November. Ahead of these meetings, Freeman again reiterated to Pym her view that the marriage bar should be abolished. Following discussions with her opposite numbers in various organisations both in the UK and in America, she had concluded that the removal of the bar was 'more in line with the Corporation's policy regarding all other women staff matters'.[114] Pym was also now of the view that the bar should be ended, commenting 'that it probably does more harm than good'.

---

[110] Same file. Pym responded with the information that Nicolls did not consider it 'an appropriate time to raise the general question of the marriage bar'.

[111] BBC/WAC:R49/857: Staff Policy Trade Unions: AWCS 1936–38, Freeman to Pym, 17 December 1937.

[112] Oram, *Women Teachers*; Glew, 'Women in the GPO'.

[113] *Ariel*, October 1937.

[114] MWP:2, Freeman to Pym, 2 November 1937. Handwritten note added by Pym.

A former Chief Inspector of Schools for the London County Council, Pym may have been influenced by the ending of the LCC bar for teachers in 1935.[115]

Despite the misgivings of Pym and Freeman, the Control Board advised the Board of Governors to tighten the rule rather than abolish it.[116] The governors, however, were themselves divided on the issue. At their meeting on 10 November 1937 Hamilton was one of two governors who spoke against the rule, although why she had changed her mind is unclear.[117] Ultimately the vote went in favour of the Control Board and the BBC marriage bar was tightened; brought 'into line with the Civil Service on all points'. On 16 November, women staff were informed that henceforth only those on Grade C and above (those earning salaries of more than £400 per annum) could be considered for retention.[118]

Yet, though the BBC marriage bar had been tightened it continued to be flouted. During 1938, the eight-month rule which had been introduced for temporary staff was waived for married women telephonists who were considered essential to the new telephone enquiry service.[119] It was agreed that married women wardrobe assistants could be employed in television as 'it is a normal custom for them to get outside employment'.[120] Following an impassioned flurry of memos from the Catering Manager, married women waitresses also became eligible for work at the BBC.[121] This was later extended to all women catering staff.[122] Ruth Field, an Assistant Producer in the Schools Department, was allowed to remain on the staff following the announcement of her marriage in November 1938, even though she earned less than £400 a year. This was because of

[115] The Open Door Council and Six Point Group had worked with the National Union of Women Teachers to pressurise the LCC to end the bar. The Women's Library/Open Door Council: 5ODC/A/05, Annual Reports.

[116] MWP:2, Note on Proposed Marriage Bar, prepared for the Board of Governors, 5 November 1937.

[117] BBC/WAC:R1/5/1/ Board of Governor's Minutes, 10 November 1937. Sir Ian Fraser also wanted the rule abolished. H.A.L. Fisher, Dr J.J. Mallon and C.H.G. Millis were in favour of retaining only senior women. The views of the other two governors (Viscountess Bridgman and R.C. Norman) were not recorded.

[118] MWP:2, Nicolls to Pym, 17 November 1937.

[119] MWP:2, Freeman to Clarke, 27 April 1938.

[120] MWP:2, Pym to Clarke, 1 March 1938.

[121] MWP:2, Clarke to Wade, 12 October 1938.

[122] MWP:2, Pym to Wade, 2 November 1938.

its 'convenience'.[123] In October 1938, Pym suggested to Lochhead (who had replaced Nicolls as Controller (Administration) in April 1938), that there should be a less rigid interpretation of the rule and the reintroduction of an element of compassion for women currently on the BBC staff.[124] Pym also re-emphasised that the 'right policy' was the removal of the bar, and he had new ammunition. Attached to his memo was a report from Ray Strachey. In the summer of 1938 Freeman, in her quest for ammunition against the BBC bar, had contacted Strachey in her capacity as Secretary to the Women's Employment Federation, asking for information about marriage bars in the UK. Strachey responded with a fulsome account of occupations that did and did not require resignation, ending with an expression of her own belief that it should be left to the women themselves to decide.[125] Strachey's report, however, failed to convince Lochhead.

In June 1938 John Reith left the BBC to take up a new post at Imperial Airways. His replacement as Director General, Frederick Ogilvie, while not seeing the marriage bar as a priority was prepared to reconsider the issue. Possibly his previous role as an academic (he had been Vice-Chancellor of Queen's University Belfast) made him more amenable to change and he intimated that he would consider a review, but not until October 1939.[126] But by then the Second World War had begun. Seeking urgent clarification about the situation for married women, Freeman was told that the marriage bar would be relaxed but only for women in Categories A and B (women who continued to work for the Corporation), it remained in force for woman in Category C (those seconded to areas of work outside the BBC).[127]

On 2 October 1939 the new ruling was promulgated to women staff who were informed that, should they marry during the war, they could remain with the Corporation but would be required to resign at the ending of hostilities.[128] In April 1940, it was reported that 43 women clerical staff had since married. The main issues the BBC now faced being whether a married secretary could work in the same office as her husband (it was

---

[123] BBC/WAC:L1/1257/3: Ruth Steele Staff File, Milliken to Rose-Troup, 25 November 1938.

[124] MWP:2, Pym to Lochhead, 3 October 1938.

[125] MWP:2, Strachey to Freeman, 20 July 1938.

[126] MWP:2, Graves to Lochhead, 12 October 1938.

[127] R49/371/3: Married Women Policy: 3, August 1939–46 (hereafter MWP:3), Pym to Freeman, 30 August 1939.

[128] MWP:3, Instruction No. 8: Employment of Married Women, 2 October 1939.

not allowed) and whether staff details should be altered to reflect their married name (only if the woman herself requested this).[129] The war-time removal of the BBC bar was in line with most occupations including the Civil Service and teaching. Under the National Service Act of December 1941, single women aged 20 to 30 were conscripted; however, married women were strongly encouraged to work albeit with exemptions for mothers of infants and school-aged children.[130]

In September 1941 Miss Freeman was redeployed as Staff Welfare Officer.[131] Her replacement was Gladys Burlton, a well-known staffing expert, who was soon questioning the differing treatment for women in Category C'.[132] Although only 24 women were affected she could discern 'no possible reason' for the continuing discrimination.[133] In February 1942, the wider issue of the BBC marriage bar was raised; Burlton scathing of the 'surprising old Corporation policy'.[134] Spurred on to prepare a report, her impassioned six-page critique of the Corporation's marriage bar was submitted on 17 April 1942.[135]

Burlton identified two major contradictions of the BBC bar: firstly that it functioned despite many aspects of it being disliked by managers and secondly its regressive nature in an institution that held enlightened attitudes towards women staff. Exposing the paucity of theories behind marriage bars, she in turn took to task the domestic argument, the financial argument and the efficiency argument producing a stark indictment of BBC policy. In considering the domestic argument, she contested whether any employer, however wise, was better fitted to decide if any individual woman should stay than the woman herself. The assumption seemed to be, 'that a girl who has always proved herself a level-headed, capable person, fit to hold a responsible position, loses all her sense immediately she marries

[129] MWP:3 General Business Manager, 1 April 1940; Pym to Freeman and Clarke, 13 June 1941; Assistant Director of Staff Administration to Assistant Head of Staff Records, 1 January 1942.
[130] Gail Braybon and Penny Summerfield (1987) *Out of the Cage: Women's Experiences of Two World Wars* (London: Pandora) pp. 157–60.
[131] Freeman's card index file shows she resigned on 30 April 1943, aged 42. In brackets under her name is written 'Mrs Ivin' indicating that she had married.
[132] After a long career in retail Burlton established, in the mid-1920s, the highly successful Burtlon Staff Agency and the Burlton Institute. She had also written many books.
[133] MWP:3, Read (Assistant Women's Establishment Officer) to Burlton, 19 January 1942.
[134] MWP:3, Burlton to Cameron, (Assistant Director of Staff Administration),12 February 1942.
[135] MWP:3, Burlton to Cameron, 17 April 1942.

and becomes incapable of judging how to conduct her life'. The principle behind the financial argument, that no-one should be allowed to work who did not 'need' to do so, logically denoted that 'no-one (man or woman) with sufficient private means should be allowed to follow a profession'. As to the efficiency argument, Burtlon declared, 'The idea that married women are as a class less efficient than unmarried women is demonstrably untrue.' She concluded with an emotional flourish:

> Why should we class marriage with misdemeanour, inefficiency, ill health and old age as a reason for dismissing a woman from her employment? Why should a married woman who has devoted the whole of her single life to mastering a profession be debarred from continuing to practise it? This is surely a grave infringement of the rights of women in a democratic country.

Burlton's fervid words had immediate effect. Two weeks later, on 29 April 1942, the new Director General, Robert Foot, agreed to rescind the ruling governing women in Category C.[136] It was also agreed that the question of married women's employment would be reviewed once hostilities were over.[137]

But while Burlton may have exposed the futility of marriage bar ideology, it was economic reality that ultimately led to the ending of the bar. In November 1943, Foot expressed the view that after the war, bars imposed by individual employers were no longer likely to be effective as married women so terminated would simply seek work elsewhere. His inclination was, therefore, that following the ending of hostilities, the BBC bar should be removed.[138] In the event, the decision came earlier. On 21 September 1944, the Board of Governors agreed that the marriage bar should not be re-imposed after the war, though this was clarified by the proviso that this would be reconsidered if, at any time, the continued employment of married women appeared to the Corporation to be against public policy.[139] Following this decree, there is no further mention of 'Married Women Policy' in the files.

---

[136] MWP:3, Pym to Burlton, 29 April 1942. Ogilvie had resigned in January 1942, his replacement a diarchy of Robert Foot and Cecil Graves.

[137] MWP:3, Beadle to Foot, 28 April 1942.

[138] MWP:3, Pym to Ashbridge, 12 September 1944.

[139] MWP:3, Clerk to the Board to Pym, 22 September 1944. Economic imperatives also saw the lifting of the marriage bar in teaching (1944) and the Civil Service (1946) where staffing shortages also made it expedient to continue the employment of married women.

4.8 Conclusion

Gladys Burlton's analysis of the inefficiencies and contradictions of the BBC marriage bar revealed what had long been the Corporation's unease with its policy: it was never convinced that married women should not work. Whereas the interwar narrative of the marriage bar in professions such as teaching and the Civil Service was predominantly that of women's battle to overturn it, at the BBC it was the story of the Corporation's justification in applying it. The self-serving nature of the BBC meant it was reluctant to dismiss valuable staff, especially those it would be difficult to replace. Added to this was the empathy of BBC administrators towards married women's appetite for work, both for financial reasons and for the fulfilment of a career. By retaining married women the BBC accepted that it was possible to 'serve two masters'; both to care for a husband and family and to remain a productive member of staff. This is important because it countered directly the widely held maxim that a woman's place was in the home.

However, the identification within the BBC of two distinct classes of married women, those who wanted a career and those whose 'heart was in the home', was in many ways a truism. In line with the convention of the day, the majority of the Corporation's female staff left voluntarily when they wed. For the few who came before the Marriage Tribunal, there was a palpable sense of discomfort amongst managers that they should be passing judgement on an individual's choice about work and domestic life, a sentiment not shared by the impersonal and monolithic Civil Service. Even though this was the model to which the BBC marriage bar would ultimately conform, the 'intrinsic needs' of the Corporation continued to include married women whose work was seen as vital to the smooth running of the organisation.

Whereas established professions such as the Civil Service and teaching displayed long-held and deeply entrenched prejudice towards women, the young BBC prided itself on being enlightened in its approach to female staff. The marriage bar, however, exposed a major inconsistency; because it was discriminatory it was out of line with the Corporation's ostensible policy of equality in terms of women employees. The tension lay in the reasons for its introduction: the BBC saw in its marriage bar a quick fix response both to the economic climate and to the issue of 'inconvenient' girls. In none of the discussions leading up to the start of the bar was the issue of inequality raised. Rather, the bar was seen as a natural development: if they operated within prominent institutions such as ICI, Prudential

Assurance, the LCC and Marconi, then surely it made sense for the BBC also to adopt the practice. By doing so, they were conforming to a convention that was accepted by most professions and industries. However, while on the one hand there was confidence that, in terms of the establishment, the BBC would be seen to be doing the right thing, on the other hand there was awareness that public opinion might view the introduction of a bar differently.

How did the BBC avoid scrutiny at a time when married women's work was a critical issue of equality campaigns? It seems likely that, because it was the lower ranks who were predominantly affected by the bar rather than higher grade women, its marriage policy was kept away from the eyes of the mainly middle-class activists. Within the Civil Service and in teaching, it was these professional women who were vigorously united in protest against the bar, whereas at the BBC this was decidedly not the case. The BBC accepted that women might want to choose both marriage and a career. It was keen that those who were deemed valuable should retain their jobs, particularly if they were in the salaried grades. And it is to the BBC's salaried women we now turn.

CHAPTER 5

'New and Important Careers': Salaried Women at the BBC

'I confess that when I was first confronted by these models of modern efficiency, a large proportion of them university-trained, others who are chartered accountants or who held administrative posts during the War, I was scared.' So claimed the effervescent *Woman's Own* journalist who took her readers 'behind the scenes' to meet the 'Women at the BBC' in January 1933. Her article included brief descriptions of the work of, amongst others, Mrs Fitzgerald, the Assistant Editor of *World Radio;* Mrs Lines, who provided photographs and illustrations 'demanded by journalists the world over'; Miss Milnes, the head librarian and Miss Glasby, who adapted plays for broadcasting 'and has written some herself'.[1] The BBC's 'career women' were a frequent source of fascination for the press in the 1930s, gripped by the breadth and diversity of roles.[2] Alongside the more conventional financial, administrative and supervisory jobs were a raft of less familiar positions such as drama producer, television make-up artist, Children's Hour Organiser, accompanist, night hostess and variety auditioner. Marjorie Scott-Johnston, for example, was one of the Assistants 'of good educational qualifications'

---

[1] *Woman's Own*, 21 January 1933.
[2] For example, 'Important Women of the BBC's Big House', *The Evening News*, 30 November 1934, 'The Women at Broadcasting House', *Good Housekeeping*, August 1935, 'The Women Who Rule the Air Waves', *Radio Times*, 12 November 1937.

© The Editor(s) (if applicable) and The Author(s) 2016
K. Murphy, *Behind the Wireless*,
DOI 10.1057/978-1-137-49173-2_5

who worked with Richard Lambert, the Editor of *The Listener,* under whose direction:

> ...she ransacked the print shops and the print departments of the British and Victoria and Albert Museums, pestered picture agencies ... devoured the resources of the London Library, scoured the continental papers and collected files of likely pictures and photographs from all parts of the world.[3]

This chapter is about the BBC's salaried women, women such as Scott-Johnson, who worked with passion and vigour in jobs that required independence of thought, creativity and administrative prowess. It does not encompass women at the top, who are considered in Chapter 6; rather it is about the general body of women who held responsible positions that were important for both the maintenance and development of the BBC, a number of whom are picked out for closer scrutiny. Women made up around 15 per cent of the BBC's salaried staff in the interwar years.[4] Without comparable figures in other similar organisations it is hard to assess whether the BBC had higher numbers; however, the Civil Service, to which the BBC was often compared, had a combined proportion of female administrative, executive and higher clerical officers of around 6 per cent.[5] Because the BBC was a brand new industry it began with no set practices and few defined positions. As Elise Sprott pointed out in a 1936 London *Evening News* article about 'The Women at the BBC':

> Many women who came into broadcasting in those early days made for themselves new and important careers, and helped to build up the reputation which British broadcasting holds in the world at large.[6]

The opportunity to create and shape the jobs they held is one of the reasons why so many of the BBC's salaried women appear to have done well.

One of the perceived legacies of the First World War in the UK was an explosion in careers for women. It was not just the BBC that provided copy for newspapers and women's magazines. As Adrian Bingham had shown, the

---

[3] Richard Lambert (1940) *Ariel and All his Quality* (London: Gollanz) pp. 136–7.
[4] BBC/WAC:R62/100/1-3: Salary Information (staff) 1923–39.
[5] Guy Routh (1965) *Occupation and Pay in Great Britain, 1906–1960* (Cambridge: Cambridge University Press) p. 24.
[6] *Evening News*, 30 June 1936.

image of the career girl and the successful professional woman became one of the signifiers of modernity.[7] The 1919 Sex Disqualification (Removal) Act had opened most professions to women and alongside nursing, teaching and the Civil Service, which were the three largest employers of middle-class women, the 1931 Census recorded 394 dentists, 195 lawyers, 107 architects and 119 accountants.[8] Although these were tiny numbers they represented the possibility of entering new areas of work, a situation mirrored at the BBC. For middle-class women, following a profession had always been about more than just remuneration; it gave them a sense of identity and a purpose.[9] The scope of careers now available to the well-educated offered heightened opportunities for challenge and stimulation.

The expectation that educated girls would have a career, and perhaps a less-conventional one, was reflected in the array of advice books and books on women's work that flourished in the interwar years. In *Women's Work in Modern England* Vera Brittain included aviation, photography and scientific work; in *The Road to Success* Margaret Cole offered retail buyer, commercial artistry and property management; Ray Strachey's *Careers and Openings for Women* suggested chiropody, publishing and showroom demonstrating.[10] In *Women and a Changing Civilisation* published in 1934, Winifred Holtby included a chapter on the rights of women to work, a page of which applauded the entry of women into new occupations like oceanography, stock-broking and the BBC.[11] The BBC's early *Women's Hour* programme (see Chapter 7) also broadcast a raft of talks on less common careers such as analytical chemist, tea room manager, sports organiser, art auctioneer and house decorator.

Despite the abundance of new opportunities, the traditional professions of teaching, nursing and the Civil Service were by far the most popular

[7] Adrian Bingham (2004) *Gender, Modernity, and the Popular Press in Inter-War Britain* (Oxford: Clarendon Press) pp. 64–8.

[8] Ray Strachey (1935) *Careers and Openings for Women: A Survey of Women's Employment and a Guide for Those Seeking Work* (London: Faber and Faber) p. 45.

[9] Martha Vicinus (1985) *Independent Women: Work and Community for Single Women* (London: Virago) pp. 171–7.

[10] Vera Brittain (1928) *Women's Work in Modern England* (London: Noel Douglas); Margaret Cole, ed. (1936) *The Road to Success: 20 Essays on the Choice of Career for Women* (London: Methuen); Ray Strachey, *Careers and Openings.*

[11] Winifred Holtby (1934) *Women and a Changing Civilisation* (London: Lane and Bodley Head) p. 83. Holtby erroneously referred to Mary Agnes Hamilton as a Director of the BBC, she was in fact a BBC Governor.

career choices for aspiring working-class and middle-class young women, largely because they were established and 'safe'. Writing about business opportunities in *Careers for Girls*, Lady Rhondda showed frustration at the tendency of parents, unless very well-off or modern-minded, 'to shy at sending girls into the newer professions'.[12] Yet it was the traditional professions that were constantly berated for their discriminatory practices be it unequal pay, low pay, poor promotional opportunities or marriage bars.[13] Unquestionably, many women in these professions gained immense job satisfaction and felt pride in the work they performed and many had long and happy careers (and happy short careers, as most would have left on marriage) but the BBC offered the possibility of a job that could not only be satisfying in the long term but was also ostensibly on equal terms with men. The Control Board minutes for 16 November 1926 are unambiguous; in a discussion on the 'Position of Women' it was agreed that 'the principle of eligibility of women for any posts must be maintained'.[14] Hilda Matheson also professed her belief that 'equal pay for equal work' was on the whole respected.[15] This notion of equality was replicated in other 'modern' professions where women worked alongside men. For example, at the John Lewis Partnership, trainee managers received the same pay for the job and there was a minimum National Union of Journalists (NUJ) rate for all newspaper reporters, regardless of sex.[16]

Mary Agnes Hamilton's claim that in broadcasting, 'men and women work on a genuine basis of equal and common concern' was widely held to be true, although, as this chapter will show, in reality 'hidden' inequalities in recruitment, pay and promotion were widespread.[17] However, there was a very real outward sense that, within the salaried grades at least, women and men at the BBC worked equally, side by side.

[12] J.A.R. Cairns (1928) *Careers for Girls* (London: Hutchinson) p. 54. See also Angela V. John (2013) *Turning the Tide: The Life of Lady Rhondda* (Cardigan: Parthian) pp. 276–8.

[13] See Helen Glew (2009) 'Women's Employment in the General Post Office, 1914–1939' (Unpublished doctoral dissertation: University of London); Alison Oram (1996) *Women Teachers and Feminist Politics 1900–1939* (Manchester: Manchester University Press).

[14] BBC/WAC:R3/3/2: Control Board Minutes, 1926, 16 November 1926. This was reiterated by Goldsmith in 1928, 'The principle of women working with equal status is accepted', 27 November 1928.

[15] *Women's Leader and Common Cause*, 2 January 1931.

[16] At John Lewis these female employees still faced indirect discrimination. Judy Faraday (2009) 'A Kind of Superior Hobby: Women Managers in the John Lewis Partnership 1918–1950' (Unpublished MPhil dissertation: University of Wolverhampton) p. 38.

[17] *Radio Times*, 16 November 1934, 'Women's Broadcasting Number'.

5.1 BEING SALARIED AT THE BBC

Although many of the BBC's waged women would have seen their work in terms of a career, it was within the salaried ranks that this was fully acknowledged. As a member of the 'officer' class conditions of employment were superior, earnings increased far more rapidly and the possibilities of mobility, advancement and autonomy were greatly enhanced. A handful of salaried women worked in areas of work that were designated 'Essentially a Woman's Job' such as the Registry and Duplicating Supervisors, the Women's Staff Administrator and her Assistant, the Matron and the Women's Press Representative (all held by women who had started their BBC careers in the waged ranks).[18] Other positions saw women predominate, such as the role of Children's Hour Organiser and those of the Photographic and Display Section. In other departments, such as School Broadcasting and Publications, men and women worked alongside. None of this is surprising at a time when work connected with children and visual arts was considered suitable female employment. However, a significant number of women worked in sections where they were either isolated (or perhaps with one other woman) in jobs that could equally, or more usually, have been held by men; such as a Talks Assistant, a Variety Producer, in Music Contracts or as a News Librarian.

Whether male or female, the lowest starting rate for a salaried position at the BBC was usually £260 per annum, this at a time when £250 was viewed as the minimum necessary for a middle-class life-style.[19] As Virginia Woolf emphasised in *Three Guineas*, £250 was seen as a good salary for women.[20] Quite a few BBC women earned upwards of £500, which gained them the enhanced perks of senior status. Although there were often complaints about the low salaries paid to BBC men, those of women compared well to other interwar careers.[21] In 1934, the Junior Executive grades of the Civil Service, where most salaried women were clustered, offered in

---

[18] BBC/WAC:R49/231/1: Grades and Salaries, Monthly (Except Grade 'D'), 20 January 1937.

[19] Ross McKibbin (1998) *Classes and Cultures: England 1918–1951* (Oxford: Oxford University Press) p. 44.

[20] Virginia Woolf (1938/2000) *Three Guineas* (Oxford: Oxford University Press) p. 217.

[21] The Corporation did not pay excessive salaries. Reith expected staff to be motivated by public service, not high pay. In comparison with barristers, doctors, dentists and civil servants, most senior BBC male employees earned modestly.

the region of £152–£396 per year.[22] The average woman teacher's salary was £265 per annum while nurses earned far lower. In 1937 the maximum salary paid by a local authority hospital was £63.17s a year.[23] More generous salaries were recorded in 'newer' professions. For example, a buyer in a West End store earned between £400 and £600 a year whilst an advertising copywriter could earn up to £750.[24]

As was the case with its waged-women staff, the unconventional nature of the BBC set it apart from most traditional professions in terms of recruitment. Teaching, the Civil Service and nursing all had formal entry requirements with young women entering straight from school, training college or university. The BBC was more akin to journalism with 'a thousand and one ways' of learning the game, or advertising which a woman could 'enter at many points'.[25] The John Lewis Partnership took a considered decision, in 1918, to employ experienced women. Under its Learnership Scheme, women from the theatre, the arts and even archaeology were enticed to work at the store where their diverse backgrounds were seen as an enhancement to the sales team.[26] Likewise at the BBC, it was unusual to be recruited directly to a salaried position without some form of wider work experience so, for instance, Olive Shapley had been a Workers' Educational Association (WEA) lecturer and trainee nursery teacher at the Rachel Macmillan School; Janet Quigley had worked in publishing and for the Empire Marketing Board and Ursula Eason had held a secretarial position at the Times Book Club.

Both Quigley and Eason had come to the BBC through connections. Quigley had been proposed as the perfect colleague by her flatmate Isa Benzie (then an Assistant in the Foreign Department). Eason (who was recruited as Northern Ireland Children's Hour Organiser) was supported by her cousin, the Variety Producer, C.H. Brewer.[27] Equally, Isabel Shields,

[22] Ray Strachey, *Careers and Openings*, p. 216; Dorothy Evans (1934) *Women and the Civil Service* (London: Pitman) p. 133. In 1934, 744 women were employed in the Civil Service Executive Grades.

[23] Routh, *Occupation and Pay*, p. 69. Brian Abel-Smith (1960) *A History of the Nursing Profession* (London: Heinemann) p. 276.

[24] Cole, *The Road to Success*, p. 241; Vyrnwy Biscoe (1932) *300 Careers for Women* (London: Lovat Dickson) p. 18.

[25] Cole, *The Road to Success*, pp. 115, 201.

[26] Faraday, 'A Kind of Superior Hobby', pp. 31–3, 51–2.

[27] BBC/WAC:L1/784/1: Janet Quigley Staff File, Anderson to Carpendale, 18 December 1929; BBC/WAC: L1/2142/1: Ursula Eason Staff File, Brewer to Clarke, 13 November 1933.

who joined the BBC as Reith's secretary in January 1923, had been suggested by Frances Stevenson, private secretary to Lloyd George.[28] As
well as commendation, the two other main ways that the BBC recruited
to salaried posts was from amongst those who wrote to the Company/
Corporation and through word of mouth. John Reith might even put in a
good word for selected individuals; for example, he recorded in his diary
'Sir Samuel Instone to see me about a job for his daughter'. Anna Instone
joined the Recorded Programmes Department in December 1933.[29]

In 1934, following an independent report into the recruitment of staff
(which noted 'a good proportion of women to men on the staff') it was
agreed that Appointment Boards should be introduced and advertising
more widely used. Prior to the mid-1930s, advertisements were only used
for specialised positions. This was because of concerns that the glamour
ascribed to the organisation could result in hundreds, if not thousands,
of applications, all of which would have to be scrutinised and sorted.[30]
Advertising, however, was seen as especially important to reach the 'full
potential of female applicants' who were generally 'so few and so scattered'
they could only effectively be reached this way. One of those who responded
to an advertisement was Olive Shapley whose mother had seen the job of
Northern Children's Hour Organiser advertised in the *Daily Telegraph*.[31]
Shapley recalled that she had worn a double-breasted navy blue gabardine
coat with a navy blue tricorn hat to the interview, which gave her future
boss the impression that she was a midwife, but nevertheless she got the job.

The introduction of Appointment Boards in May 1934 may actually
have hindered women's entry into salaried positions at the BBC but even
if they did not necessarily improve women's chances of getting in, they did
establish an element of transparency.[32]

The Staff Training School, established in October 1936, was another
way of introducing an element of consistency into the BBC's recruitment
and promotion procedures. While it was expected that new arrivals to
the salaried staff would attend, 'veterans' were also encouraged to apply.
The 12-week course, run internally by the BBC, gave an overview of the
Corporation's organisation and activities; offered opportunities to study

---

[28] Reith Diaries, 2 January 1923.
[29] Reith Diaries, 19 January 1933.
[30] BBC/WAC:R49/31/1: Report on Recruitment of Staff 1934 by D.B. Mair and Ernest
Barker.
[31] Olive Shapley (1996) *Broadcasting: A Life* (London: Scarlet Press) p. 33.
[32] See BBC/WAC:R49/27/1-3: Appointment Boards: Minutes.

programme, administrative and engineering practice and tested the abilities of staff seeking advancement.[33] The twenty staff who attended the Summer Course in 1939, for instance, included four women. Amongst them were Kathleen Lines, the long-serving Head of the Photographic Section, but also Eileen Molony and Gwen Parry Jones, both newly recruited as Children's Hour Organisers for Bristol and Cardiff respectively.

Miss Freeman, the WSA, sat on Appointment Boards when women were being interviewed, occasionally chairing the session. Although her role at the BBC was predominantly associated with the weekly paid grades, she certainly championed at least two separate initiatives to improve the recruitment of salaried female staff. One was by making links with the Women's Appointment Boards at upward of fifteen universities which included Oxford and Cambridge, London, Birmingham, Liverpool, Edinburgh and Aberdeen.[34] Ursula Eason, for instance, was notified by the University of London about the *Children's Hour* vacancy.[35] Another was by affiliating, in May 1935, to the Women's Employment Federation which operated as a clearing-house for employers.[36] On at least three occasions they were approached by the BBC with requests for assistance in securing senior female staff.[37]

The BBC's salaried women were widely dispersed within Administration, Programmes and Public Relations, sometimes in traditional roles, often in positions that had been created or developed specifically for the new medium of broadcasting. The one division in which they were not utilised was Engineering. Areas of the BBC that, post-war, would become increasingly hostile to salaried women, such as News, Outside Broadcasts and Light Entertainment, did employ small numbers. The careers of the Corporation's four highest-earning women, Hilda Matheson, Mary Somerville, Isa Benzie and Mary Adams are explored in the next chapter. Here, the spotlight is thrown on a selection of women who also held significant roles, offering brief insights into their BBC work.

---

[33] BBC/WAC:R49/709/1: Internal Instruction 415, Staff Training Department, 24 November 1937.

[34] BBC/WAC:R1/69/3: Board of Governors, DG's Reports and Papers, Notes on Procedure in Regard to Staff Appointments for Submission to Barker and Mair, 8 November 1933.

[35] *Radio Pictorial*, 12 March 1937.

[36] Women's Library/Women's Employment Federation: 6/WEF/487, Executive Minutes 1933–37, 9 May 1935. WEF was founded in 1933 under the Secretaryship of Ray Strachey.

[37] For example, WEF: Exec Minutes, 13 June 1935, 8 October 1936.

Librarian

In July 1958, on the occasion of her farewell tea party, Florence Milnes was lauded by the Director General, Ian Jacob, as a 'practical visionary' who had grasped that a first-class library would be essential to the BBC.[38] As Head Librarian for more than thirty years, Milnes had indeed built up the service from scratch, her Reference Section one of a sextet of Head Office libraries that were developed in the interwar years. Libraries were fundamental to the BBC. They were the repositories of the sheet music, orchestral scores, gramophone records and play scripts that were at the core of entertainment; they housed the stockpiles of photographs that were necessary for illustration and promotion; they provided the reference books and press cuttings crucial for information and accuracy and they became the receptacle for selected sound recordings. Women were key to them all and not simply in terms of filing or cataloguing. Several were experts who managed the collections, who had ambition for their expansion and who became supremely knowledgeable in their specialisms. Kathleen Lines, like Milnes, originated her section, the Photographic Section. Anna Instone and Marie Slocombe, who joined the BBC in the mid-1930s, were in at the birth of the Gramophone Library and the Sound Archive respectively. Alice Wright, as Deputy Music Librarian and Miss Shiel as Play Librarian, also held important roles.

By the 1930s librarianship had become an acceptable career for women, with the possibility of advancement via a Diploma, although appointments to senior posts were still rare.[39] The libraries at the BBC were, however, far from conventional. None of the BBC women had been schooled in librarianship although Anna Instone had trained at the Royal College of Music and Kathleen Lines had purportedly run a commercial arts studio. All but Lines joined the BBC as weekly paid staff with no indication that they would progress to significant roles.

According to Florence Milnes, it was her transference to the quickly conceived News Unit during the 1926 General Strike that convinced her that the BBC needed a proper library.[40] She had joined the Company the previous year, aged thirty two, as a £3.10s a week Information Assistant, much of her job gathering background research for programmes. In April 1927, she was given full charge of the Library, along with a pay rise and

[38] BBC/WAC:L1/705/1: Florence Milnes Staff File, Ian Jacob speech, 24 July 1958.
[39] Strachey, Careers and Openings, p. 228.
[40] The Library World, 1959, pp. 171–5.

promotion to the salaried grades. What began as a handful of books and loose press-cuttings had, by 1932, grown into a collection of around 1500 volumes; the press-cuttings now pasted into large card-indexed albums for ease of access.[41] The purpose-built facilities at Broadcasting House were widely used. The *Radio Times* in 1934 was impressed by the scope of visitors: a member of its staff checking a last-minute reference; a producer looking up historical details for a period play; someone from the Talks department ransacking the poetry shelves.[42] Milnes spent much of her day answering queries, a light-hearted *Radio Pictorial* article revealing a stream of oddities she was asked.[43] She was a BBC 'personality', highly distinctive with her trademark suit and Eton crop. Evidently formidable, she was an awkward woman. Her annual reports expose not only her zeal but her bad manners, her cryptic ways, her temper, and her disgruntlement.[44] Milnes believed she was underpaid and her 1939 salary of £525 compared poorly with other women in similarly responsible BBC roles.

Kathleen Lines, conversely, earned £720 in 1939. Two years older than Milnes, she joined the BBC in 1924 as secretary to the Director of Education, Stobart.[45] Amongst her assorted jobs prior to the BBC were (according the newspapers) accountancy, work on an Australian pineapple farm and managing a sizeable munitions department during the war.[46] How she became responsible for photographs is unclear but by April 1925 she had assumed this role. Her Photographic Section, as it came to be known, provided photographs and illustrations for BBC publications such as *Radio Times* and *Ariel*, for its yearbooks and handbooks and for the pamphlets that supported Talks programmes, Adult Education and School Broadcasts. It also supplied national and international newspapers and journals with images of the BBC: from posed staff portraits, to studio action shots, to 'all the complicated gear at the transmitter stations'.[47]

[41] BBC/WAC:R13/301: Secretariat: Library.

[42] *Radio Times*, 14 September 1934.

[43] *Radio Pictorial*, 13 March 1936.

[44] The creation of a separate News Information Library in 1934 under Horatio Batchelor, who had held a similar post at *The Times*, was a source of great disappointment to Milnes. The Oral History of the BBC: Elizabeth Barker interview, 9 May 1983. Barker was recruited to assist Batchelor.

[45] Reith Diaries, 15 March, 1924.

[46] *Evening News*, 30 November 1934; *Woman's Own*, 21 January 1933; *Daily Dispatch*, 2 March 1933.

[47] *Radio Pictorial*, 21 December 1934.

The rich pictorial record of the early BBC that exists today is largely because Lines arranged for specific photographs to be taken. By 1933 the Section housed an estimated 6,000 images. Lines worked closely with Richard Lambert, Editor of *The Listener* who saw her as integral to the paper's success. He believed that as a large, well administered and self-supporting section, it would 'long ago have been elevated to the deserved dignity of a Department' had it not been largely staffed by women.[48] It is impossible to know if Lambert's assertion was correct because of the confused nature of the BBC's management structure, but he nevertheless sensed an injustice.

Anna Instone and Marie Slocombe, who would come to hold substantial positions within the BBC (as Head of Gramophone Programmes and Sound Archives Librarian) were, in the mid-1930s, at the start of their lengthy careers. Instone arrived in the brand new Sound Record Section in 1933, as a Gramophone Programmes Assistant, Slocombe as a summer-relief secretary in Recorded Programmes in 1937.[49] Already by 1939, both women were committed to raising the profile of their sections. Slocombe had quickly realised the importance of preserving the BBC's sound recordings, helping to accumulate 2,000 discs including the voices of H.G. Wells, George Bernard Shaw, Churchill and J.B. Priestley. The year after Instone's arrival had seen a sizeable jump in the use of gramophone records on the BBC. Although not directly a 'librarian', as usage grew she was part of the small team that contemplated with how these new programmes should be developed.[50] By 1936, the *Morning Post* was reporting on the 50,000 gramophone records in her library, an index system assisting the onerous task of compiling programmes on a diversity of themes.[51]

To work in a BBC library was an important role and one with a far wider brief than was usually associated with the position. While some activities would have been recognisable in the 1920s and 30s, such as the functions of

[48] Lambert, *Ariel and All his Quality*, p. 137. In 1937, the staff of 16 included five salaried Assistants, five shorthand typists, five clerks and a 'boy': just two of the clerks were men. BBC Staff List, 1937.

[49] BBC/WAC:L1/2,160/1, Anna Instone Staff File; Marie Slocombe: Sound Archives Librarian, available at: http://www.bbc.co.uk/archive/archive_pioneers/6501.shtml (accessed 10 March 2015).

[50] Asa Briggs (1965) *The Golden Age of Wireless: The History of Broadcasting in the United Kingdom*, Vol. 2 (London: Oxford University Press) p. 36.

[51] *Morning Post*, 18 May 1936.

the Reference Library and the Photographic Section (something similar to which would have existed within newspapers) other areas such as Recorded Sound and Gramophones were innovative and new. A number of men were employed in this BBC work, most notably Frank Hook who ran the Music Library from the start. But these were roles largely grabbed and expanded by determined women, who saw the potential to make their mark.

Accompanist

Amongst the posed photographs that survive from Kathleen Lines' collections are several of Cecil Dixon (that is the correct spelling of her name), the BBC's first staff accompanist. Music was at the heart of the BBC; three-quarters of what was broadcast.[52] Almost all musicians were on contract but a handful were staff 'accompanists', pianists who could play all styles of music and who were available to 'fill-in' at any given time. This was particularly pertinent in the early days when programme planning was in its infancy and long interludes on air were frequent. Around a dozen individuals would work at Head Office in this capacity during the interwar years, while a handful of others were employed in the regions. Three London-based women, Cecil Dixon, Jean Melville and Doris Arnold, would become celebrated radio stars.

Dixon's 21-year career on the BBC staff began in January 1923, making her one of the earliest employees. Her talent was spotted by Stanton Jeffries (the future London Station Director) at the Royal College of Music and she joined the BBC 'as a joke'.[53] One of her first roles was as 'Aunt Sophie' on *Children's Hour*, accompanying the antics of 'Uncle Arthur' (Arthur Burrows), 'Uncle Rex' (Rex Palmer) and 'Uncle Caractacus' (Cecil Lewis), and she would continue to be the pianist for the London programme for many years. She was also a noted recitalist in her own right and was regularly billed in the daytime and evening schedules as a soloist. As well as her own solo concerts, Dixon's work included playing at BBC auditions, taking part in rehearsals and accompanying soloists and musicians of national and international repute. She remained with the BBC until 1945 after which time she performed occasionally as a guest artiste.

Jean Melville, an alumna of the Royal Academy of Music, joined the BBC as an accompanist in 1927, following occasional appearances on

---

[52] Briggs, *Golden Age of Wireless*, p. 35.
[53] *Radio Times*, 14 December 1934.

contract. Once on the staff she played regularly with Jack Payne and the BBC Dance Orchestra (in the late 1920s, at the pinnacle of their BBC success) and then in the Vaudeville and Variety concerts that followed. She was well known for her synchronised piano playing, her opposite number always a man. Following her resignation in 1936 (because, having married, she 'wanted a bigger salary'), *Radio Pictorial* ran Melville's 'BBC from the Inside' series over four consecutive weeks.[54] Here she gave a glimpse of her job, in particular her pivotal role in auditions which she estimated amounted to 25,000 over the nine years. The extent of Melville's popularity was evident from her fan mail; she sent out 'thousands of photographs' each year. Although no longer on the BBC staff, Melville continued to make sporadic appearances in the late 1930s and during the war, returning as a regular on the post-war programme, *Workers' Playtime.*

Melville claimed to have played a crucial role in Doris Arnold's promotion to staff accompanist in 1928.[55] Arnold's 'rags to riches' story was endlessly retold in the press and indeed, from a £2 5s a week shorthand typist in the Stores Department she would rise to be one of the highest paid BBC women of the interwar years.[56] Arnold was a talented pianist who, encouraged by her managers, was trialled as a BBC accompanist.[57] In 1933 her duties were extended to include musical arrangements for series such as *Songs from the Shows, Music Hall* and the *Kentucky Minstrels.* She also teamed up with a colleague, Harry S. Pepper, and the pair soon became famous for their double piano playing (they would marry in 1943). From October 1937, Arnold produced and presented *The Melody Is There,* followed a year later by *These You Have Loved.* Compiled of extracts from light and classical music gramophone records, the programmes cemented Arnold's status as a radio star and established her as the UK's first female disc jockey. *These You Have Loved* continued to be hosted by Arnold until 1963, the last 11 years on contract to the BBC. The intensity of Arnold's workload was recognised by her managers. One commented, in 1936, that if she should fall ill, it would require three people to replace her and even then their work 'would not be equal' to hers.[58] Despite this,

---

[54] *Radio Pictorial,* 25 June 1937 to 16 July 1937.

[55] *Radio Pictorial,* 25 June 1937. Melville's assertion was that, because she was too busy to work on the orchestral arrangements for the BBC Wireless Chorus and Orchestra, Arnold had taken her place.

[56] *Oxford Dictionary of National Biography,* Doris Arnold, entry 105932 by Kate Murphy.

[57] BBC/WAC:L1/15/1 Doris Arnold Staff File (forthwith DAF), Ad.Ex. to Graves, 2 April 1928; Prod.Ex. to Ad. Ex., 16 April 1928.

[58] DAF, Variety Ex. to Beadle, 13 January 1936.

she had to push to get her salary raised to a comparable rate with her male colleagues.[59] It was also eventually acknowledged that, as a woman, her expenses were greater because she was always expected to be stylishly attired (radio concerts in front of an audience required evening dress).[60]

The salaries of the BBC's women accompanists had been remarked upon from the start. In 1925 the Programme Board discussed Dixon's inadequate remuneration. A man performing similar duties, it was agreed, would be paid at least twice the amount and her salary was immediately raised by £119.[61] Arnold's salary of £660 in 1939, while above that of several male colleagues, was still considerably less than Berkeley Mason (on £850) and Ernest Lush (on £720). Jean Melville's dissatisfaction with her salary is also understandable; the £400 she earned when she left had not risen for five years. Neither was Dixon's increased after 1934, when she reached her grade 'roof' of £600. That women should be employed as accompanists by the BBC is unsurprising, piano playing was a noted feminine art. What is more unexpected, perhaps, is that they were pro-portionally so few. Melville and Arnold did stand out. The ubiquitous nature of broadcasting saw them thrust into the limelight and both gained the status of glitzy radio stars. Dixon, although well known, was never as dazzling as her colleagues. Her main association remained with *Children's Hour.*

Children's Hour Organiser

Children's Hour was one of the BBC's first regular programmes, broad-cast each weekday from December 1922.[62] Locally produced, the ear-liest editions were amateur affairs put together by station managers in the guise of 'Uncles'. However, at the close of 1923 each provincial sta-tion was instructed to appoint 'one good woman of personality, educa-tion and standing' to supervise programmes for women and children. Dorothea Barcroft in Birmingham, Ruby Barlow in Nottingham and

---

[59] DAF, Arnold to Clarke, 12 September 1933; Clarke to Ed.Ex., 30 September 1933.
[60] DAF, Arnold to Freeman, 19 July 1935.
[61] BBC/WAC:R34/600/02: Programme Board Minutes, File 2, 19 January 1925. Salary Information.
[62] The history of Children's Hour is told in Wallace Grevatt (1988) *BBC Children's Hour: A Celebration of Those Magical Years* (Lewes: The Book Guild). See also Asa Briggs (1961) *The Birth of Broadcasting, The History of Broadcasting in the United Kingdom* Vol. 1 (London: Oxford University Press) pp. 253–4, 258–62.

Kathleen Garscadden in Edinburgh were amongst those recruited, all of whom spent many years with the BBC.[63] By the mid-1920s, apart from in London, *Children's Hour* would become the preserve of women staff. In an era of social maternalism, when women's role as protectors and educators of children was venerated, the BBC viewed the job as ideal for educated women who, it was believed, were best placed to understand children's needs.[64]

From the mid-1930s the specific post of Children's Hour Organiser was created. One of the earliest recruits was Oxford-educated Olive Shapley who took up her post in Manchester in 1934. In her autobiography she vividly recreated the frenetic nature of the job which included overseeing at least two full-length plays a week, negotiating musical items with the BBC Northern Orchestra, organising monthly children's auditions as well as producing competitions, poetry readings and broadcast talks.[65] Shapley also recalled that the first advice she was given, by another woman staff member, was to learn how 'the gentlemen like their tea'.[66] In June 1935, Ruth Field, newly appointed as Children's Hour Organiser for the Midland Region, contributed 'A Day in my Life' for *Radio Pictorial* which revealed her to be a graduate of Somerville College who lived at home with her family.[67] This particular day's work began with a perusal of the morning's post (supplied by her secretary). She then discussed a manuscript; organised an audition; attended a conference about plays; mused over a story with a writer and prepared for that evening's broadcast at 5.15 pm, which involved, ahead of transmission, a rehearsal with the invited contributors. It was Field who welcomed the listeners and ultimately said "Good Night Children" at 6 o'clock. The post of Children's Hour Organiser attracted high-calibre women. Christina Orr, recruited to Edinburgh in 1936, was another graduate of Somerville, her later experience as a novelist, dramatist and editor of a Scottish children's magazine making her the ideal candidate. It also enabled her to negotiate the sizeable starting salary of £500 a year.[68]

[63] BBC/WAC:CO9:BBCo: Station Directors Meetings: Minutes, 11 December 1923.
[64] See, for example, Jane Lewis (1990) *The Politics of Motherhood: Child and Maternal Welfare in England 1900–1939* (London: Croom Helm) pp. 89–102; Oram, *Women Teachers*, pp. 17–21.
[65] Shapley, *Broadcasting: A Life*, pp. 40–5.
[66] Shapley, *Broadcasting: A Life*, p. 37.
[67] *Radio Pictorial*, 21 June 1935.
[68] BBC/WAC:L1/328/1: Christine Orr Staff File, Letter of Application, 7 October 1936.

Children's Hour in London was the exception to the all-female rule. While women such as Geraldine Elliot, Eve Russell and Barbara Sleigh worked on the programme as Assistants, they were managed by men. In 1933, the legendary Derek McCulloch was appointed London Children's Hour Organiser, promoted to Director of Children's Hour in 1935, with May Jenkin his second-in-command. Why Head Office continued to appoint men to London's *Children's Hour* is unclear. There was a super-visory element to the position but this fails to explain why women were not considered. Possibly the long association of the London programme with senior men defined it as a post for male managers and the tradition subsequently continued.

For a salaried BBC woman, the position of Children's Hour Organiser was an attractive one. It was a largely autonomous role and was often the most senior post held by a woman in the Regions. It was also a position from which to aspire. Christine Orr, for example, became a Talks Assistant in 1940 and Ruth Field, a Producer in Schools Broadcasting in 1936. For others, it launched distinguished BBC careers including that of Olive Shapley who, as we shall see, went on to pioneer social documentaries and who would become a presenter of *Woman's Hour* in 1949. Ursula Eason, who came to the BBC as Children's Hour Organiser for Northern Ireland in 1933, rose to be Assistant Head of Northern Ireland Programmes in 1949 and ultimately Assistant Head of BBC Children's Programmes in 1955.

Advertising Representative

From the early twentieth century, educated women had taken to the new profession of advertising with alacrity.[69] The Association of Advertising Women had been set up in 1910 and in 1923, the Women's Advertising Club of London was established with 50 top women executives becoming members.[70] At the BBC, the launch of *Radio Times* in 1923, followed by the introduction of two new BBC periodicals, *World Radio* in 1926 and *The Listener* in 1929, meant staff were needed to sell advertising space.

---

[69] For a discussion on women advertisers in the interwar years see Fiona Hackney (2011) '"They Opened Up a Whole New World": Feminism, Modernity and the Feminine Imagination in Women's Magazines, 1919–1939' (Unpublished doctoral dissertation: University of London) pp. 72–80.

[70] Kate Murphy (2001) *Firsts: The Livewire Book of British Women Achievers* (London: The Women's Press) pp. 38–9.

The nucleus of a department had been established at Savoy Hill. In 1926 a Miss K. Lewis, listed as earning £300 for 'propaganda', probably denotes advertising. In June 1929, Hilda Prance was appointed to the newly established post of Advertising Canvasser, along with three men. The men had all left the BBC within five months but Miss Prance would remain with the Corporation, a stalwart of the department until her death in-service in 1937.[71]

Prance was educated at Cheltenham Ladies' College and after the First World War joined the advertising staff of George Newnes. Newnes published *Radio Times* and Prance asked for a transfer to the publication. At the BBC, she had responsibility for advertising that represented women's interests, her knowledge of this market seen as advantageous.[72] That women were the prime purchasers of consumer goods was well established by the mid-1920s. Ethel Wood, Director of the advertising agency Samson Clark and Co, suggested that they were the target for possibly 75 per cent of all adverts.[73] Prance may have worked on other accounts but women's interests was certainly a large part of the job, her high earnings (according to the Bonus List for 1930, her £375 salary was the third highest) indicate it was viewed as strategically important.[74] The Bonus List also showed that out of a staff of nine canvassers (eight of whom were men), Prance was one of only four to be paid a full bonus and she was described as a splendid worker who got excellent results.

According to *Ariel*, the BBC's Advertising Department functioned like a businessman's club and only male typists were used. This was because 'the work would be too strenuous and the air (occasionally) too blue' to be coped with by women.[75] In consequence, any woman who worked in the Department would have been isolated, without even female secretarial and clerical support. In March 1933, a second woman was recruited, Mrs Carvell, 'with the idea of strengthening the canvassing of women's advertising interests in our publications'.[76] Miss Cheseldine joined the team in March 1938, selected at an Appointment Board above six men.[77]

---

[71] Greenhow, Relph and Scott were appointed on salaries of £312, £400 and £400 respectively. Prance on a salary of £300. At her death she earned £475.

[72] BBC/WAC:R13/321: Advertising Department, Goldsmith to Carpendale, 6 September 1933.

[73] J.A.R. Cairns (1928) *Careers for Girls* (London: Hutchinson) p. 170.

[74] BBC/WAC:R49/10/1: Advertising Department: Grades and Bonus System.

[75] *Ariel*, June 1936.

[76] Carvell was designated Advertising Representative.

[77] R49/27/1-3: Appointment Boards, 21 March 1938.

Prance died suddenly from bronchial pneumonia in January 1937, her obituary in *Ariel* offering a glimpse into her private life. An ardent motorist who also owned a motorcycle, she was a country lover and an enthusiastic gardener who spent much of her spare time in her cottage garden at Bourne End. The accompanying photograph presents an old-fashioned woman (her hair is in a bun) dressed in jacket, collar and tie, which *Ariel* described as her 'duty clothes'. Interestingly, a photograph of her replacement, Miss Cheseldine, also shows a young woman sporting a shirt and tie, distinctively masculine clothes which suggest that Prance and Cheseldine assumed something of a male persona.[78] This may have increased their acceptability in an area of work that was still pioneering for women and which, at the BBC, was dominated by men.

Drama Producer

Two women would gain acclaim as radio drama producers in the interwar years, Barbara Burnham and Mary Hope Allen. The BBC had quickly discovered that radio drama and stage productions were two very different beasts. Without visual clues, early broadcasts 'live' from the West End were impossible to follow, rather plays and dramatisations needed to be specifically created or adapted for wireless.[79] Val Gielgud (the elder brother of actor John) became the champion of the new form and in 1933 a distinctive Drama Department was established under his leadership.[80] Peter Creswell, Laurence Gilliam, Howard Rose and Lance Sieveking all made their names here. However, unlike her illustrious male colleagues, Mary Hope Allen's producer job would be hard won.

A former student of the Slade School of Art, Allen had worked as a freelance journalist, a copy writer, a book reviewer and a drama critic before she joined the BBC in 1927 aged 28.[81] It is therefore unsurprising that she quickly became frustrated with her £3 10s a week job as a cataloguer in the Play Library.[82] While her bosses prevaricated

[78] This was not uncommon in the interwar years. See Alison Oram (2007) *'Her Husband Was a Woman!' Women's Gender-Crossing in Modern British Popular Culture* (London: Routledge).

[79] Briggs, *Birth of Broadcasting*, pp. 280–1; *BBC Handbook*, 1928, pp. 115–16.

[80] Briggs, *Golden Age of Wireless*, pp. 160–9.

[81] BBC/WAC:L1/659/1:Mary Hope Allen Staff File 1 (hereafter MHAF:1), Allen to Nicolls, 17 October 1933.

[82] MHAF:1, Jeffrey to Eckersley, 4 January 1928.

about how best to promote her, she arranged a private interview with Reith, who agreed that she should have a production job.[83] This was shortly after the founding of the Research Section which was experimenting with new ways of making radio features and drama, and to which Allen was assigned (Sieveking's ground-breaking *Kaleidoscope* was created here).[84] M.H. Allen's first production (she always used her initials) was *Russian Twilight* in 1929, described by Gielgud as having 'a distinction of taste and judgement'.[85] She remembered it as 'full of sombre harmonies, samovars, icons and philosophic peasants sitting on their doorsteps'.[86] Doris Arnold worked with her on the programme, selecting and arranging the music, which Allen then intertwined with words and sounds. Amongst other work was *The Forsaken City*, which 'startled the public' with its use of the writings of Defoe, Pepys and others to create a 'picture in sound' of the Great Plague.[87] In April 1932, the first of Allen's *Miscellany* programmes was aired, a new departure for radio. This was a collection of plays, songs and poems made in collaboration with the Variety producer, Denis Freeman. Yet despite a consensus that she had ability, it was felt at first that Allen lacked confidence (perhaps understandable in a unit where she was the only woman) although by 1932 Gielgud observed that she had 'arrived'.[88] The Research Section was eventually seen as too 'high-brow', closing in 1933 and Allen was transferred to the new Drama Department.

Allen was a productive member of the Drama team. For instance, amongst her 16 productions in 1936 were *Louisa Wants a Bicycle* (subtitled, 'The Fight for Woman's Freedom'), a race through women's emancipation which included sketches of the likes of Mary Wollstonecraft, Florence

---

[83] *The Times*, 9 April 2001.

[84] Mary Hope Allen Staff File 2, (hereafter MHAF:2), Eckersley to Reith, 14 June 1929. The brief of the Section was to find experimental new ways of making radio features and drama. See Paddy Scannell and David Cardiff (1991) *A Social History of British Broadcasting, 1922–1939* (London: Basil Blackwood) pp. 135–40. See also David Hendy (2012) 'Biography and the Emotions as a Missing "Narrative" in Media History: A Case Study of Lance Sieveking and the Early BBC', *Media History*, 18 (3–4), 361–78.

[85] MHAF:2, Gielgud to Reith, 4 February 1930.

[86] *News Chronicle*, 15 August 1939.

[87] Maurice Gorham (1948) *Sound and Fury* (London: Percival Marshall) p. 35; *Radio Times*, 26 February 1932.

[88] MHAF:2, Gielgud to Eckersley, 17 September 1929, 4 February 1930; Confidential Report February 1931, 6 September 1932, Gielgud to Carpendale.

Nightingale, Millicent Fawcett and Elizabeth Garrett Anderson.[89] *London Calling-1600*, an evocation of life in the reign of Elizabeth I, was praised by *The Listener* as 'the most enchanting, the most witty and the most original entertainment I have yet heard over the air'.[90] Gielgud similarly enjoyed the production and also commended her for Henry James' *Four Meetings* and the comedy, *Youth at the Helm*.[91] Whereas much of Allen's work was lyrical and avant-garde, that of her colleague Barbara Burnham was more traditional. Burnham excelled at adapting and producing novels and stage plays for radio.

After studying at Elsie Fogerty's Central School of Dramatic Art, Burnham had arrived at the BBC in 1930 (initially on a programme contract rather than as staff), bringing to the listener adaptations of the works of Shakespeare, Poe, and Marlow. In 1934, her first credit as a producer was for Chekhov's *The Seagull* and she produced four other of his plays. In a *Radio Pictorial* interview in 1935, Burnham explained the process of adaptation and production, how she cut, rewrote and reshaped a novel or a stage play into radio form. It was a matter of 'simplifying, orchestrating and translating scenery into sound'.[92] Her 1936 production of Euripides' *The Trojan Women* starred Flora Robson and she would direct many of the theatre greats.[93] As well as classic plays, she adapted and produced the novels of J.B. Priestley, Edgar Wallace and John Galsworthy and worked with modern playwrights; Aimee Stuart's *Aunt Jeannie* was written specifically for radio in 1938. Burnham's closest collaboration was with the novelist James Hilton, starting with an adaptation of *Goodbye Mr Chips* in 1935, followed by *The Lost Horizon, And Now Goodbye* and *We are Not Alone*. Describing the process of working together on *The Lost Horizon*, Burnham portrayed how they kept 'hurling it backwards and forwards for a few weeks' before agreeing that she 'might as well do the rest in production'.[94] On this occasion, she had read the book by chance and had contacted Hilton about it; at other times, dramas were suggested and discussed at weekly departmental meetings.

Both Barbara Burnham and Mary Hope Allen had long and successful BBC careers. Allen remained in radio for her entire career, retiring

[89] *Radio Times*, 8 September 1936.
[90] *The Listener*, 22 April 1936.
[91] Mary Hope Allen Staff File 2, Confidential Report, 1937.
[92] *Radio Pictorial*, 8 March 1935.
[93] *Radio Times*, 25 April 1937.
[94] *Radio Pictorial*, 17 July 1936.

in 1958 and Burnham was still producing BBC television plays in the early 1960s. Burnham's pre-war salary (£800 in 1939) was considerably more than Allen's (£620) as will be discussed below. Perhaps Burnham had an easier ride because she was a more traditional drama producer with a less combative personality. Intriguingly, the first dramatic interludes on the wireless were also pioneered by a woman, not within the Drama Department, but in School Broadcasting. They were to cement the career of Rhoda Power.

School Broadcaster

Rhoda Power is acknowledged as one of the outstanding talents of School Broadcasting. From her initial talk on 'Boys and Girls of the Middle Ages' in September 1927 she displayed an extraordinary gift for storytelling and the knack of bringing 'History' alive. Prior to the Second World War she was the most widely used daytime speaker on the BBC. She delivered more than 400 talks for the Schools Department and she continued to work with the department both broadcasting and later, writing scripts for broadcasting, until her death in 1957. Although not an academic historian, prior to her arrival at the BBC she had written several history books for children with her sister Eileen Power, Professor of Economic History at the LSE.[95] These provided much of the source material for her early history talks. However, Power's broadcasting career had actually begun in January 1927 when she presented a series of afternoon talks for Hilda Matheson. At some point that year, Mary Somerville approached Power about broadcasting for schools. Like all those who gave talks on the BBC, this was to be on a contract basis but Power's unique association with the Schools Department would ultimately see her become an established member of staff.[96]

Without doubt, Power's writing and presentation style were 'vivid'.[97] At first her talks were straight reads with 'one eye on the clock' to ensure she finished exactly on time.[98] From February 1928 dialogue, sounds and music were added to conjure up the atmosphere of the times. The 'eureka' moment was a broadcast about the Elizabethan Lord Mayor of London,

---

[95] For example, *Boys and Girls of History, Twenty Centuries of Travel: A Simple Survey of British History and Cities and their Stories: An Introduction to the Study of European History* were all published between 1926 and 1927.

[96] This had severe repercussions later for her pension.

[97] *Radio Times*, 16 January 1928.

[98] *BBC Year Book*, 1950.

Edward Osborne. At the rehearsal, as Power described the crowds cross-ing London Bridge 'the hubbub, the jostling, the singing', her producer George Dixon suggested the songs be sung, approaching and receding from the microphone to create a sense of realness.[99] By 1931, these sound-scapes had developed into full-blown 'dramatic interludes' using actors (and staff members) to recreate historical scenes. As well as expansive series on British history, Power also conceived original talks on mythology and legends. From 1929 'Stories for Younger Pupils' were added to her repertoire and in September 1931, she broadcast her first 'World History' talk.[100] From 1933, these would be presented in collaboration with her sister: Eileen Power was contracted to deliver the 'straight' talk which alternated with a corresponding 'dramatic interlude' from Rhoda.

Power's output was staggering, on many occasions she presented two talks consecutively, with just a five-minute break in between. In February 1935, *Radio Times* included a 'message to schools' in which she outlined that her part at the microphone was 'to try to make the history in these lessons come to life'. The pupils' part at the loudspeaker was 'not only to listen, but to try to see with your ears'. In this way history, Power explained, 'instead of being the "dry bones of the past", has become a living thing' revealing her deep understanding of the empathic nature of radio.[101] However, the 'interlude' technique she pioneered was to prove problematic. As she explained to Mary Somerville in May 1936, whereas the straight lesson (fee: six guineas) and the story for younger children (fee: two guineas) was material that could be turned into books, the 'inter-ludes' (fee: nine guineas), whether illustrated lessons or dramatisations were 'so radiogenic' they could not be used for any other purpose.[102] With her earnings from the BBC no longer sufficient she asked to be employed on a more solid basis and from July 1937 she was engaged on a fixed programme contract by the Schools Department, her salary estimated at £600 a year.[103]

Power was probably the only individual in the interwar years for whom broadcasting became their primary occupation. She was passionate about her educational work and it was acknowledged that she undersold herself

---

[99] 'Faith, Hope and Clarity', broadcast on 4 April 1934.

[100] *Radio Times*, 28 September 1931.

[101] *Radio Times*, 6 February 1935.

[102] BBC/WAC:R94/2,962: Rhoda Power Talks File 1 (hereafter RPF:1), Power to Somerville, 30 May 1936.

[103] RPF:1, Cruttwell to Somerville, 16 July 1937.

in order to ensure a constant flow of BBC work. For Somerville, Power's was the 'most outstanding contribution' to the Department's programmes for elementary schools, going far beyond what was usually required of broadcasters, not only scriptwriting and delivering her talks, but planning, researching, teaching and editing them too.[104]

Assistant Editor

On 7 November 1930, Janet Adam Smith wrote a letter to her boss, Richard Lambert, the Editor of *The Listener* in which she informed him that she wished to apply to be his 'second in command'. Although she realised that there 'might be a feeling against a woman' holding this position she assured Lambert that she liked responsibility and was exceedingly keen to succeed.[105] She also pointed out that '*World Radio*' had an Assistant Editor, Ella Fitzgerald. Adam Smith had worked on *The Listener* since its inception the previous year. The Assistant Editor, Moray McLaren, had just left and she felt sure that she was up to the task. Lambert was in agreement and for the next five years, until she left to be married in 1935, Adam Smith would fulfil an indispensable role on *The Listener*, not only overseeing literary reviews but bringing modern poetry and art to its pages.

It was a friendship between her father and Reith that brought Adam Smith to the BBC in 1928. An Aberdonian, she had just left Somerville College with an English degree and was looking for work in London. She started at the BBC as a £3 10s a week shorthand typist in the Publications Department but her appetite for more challenging work was soon noticed and, in May 1929, she was offered a job as Lambert's 'secretary-cum-assistant' on *The Listener*, the first edition of which had been published in January that year.[106] Very quickly she had taken on a far broader role. Working alongside McLaren, she sub-edited manuscripts, kept contact with contributors, wrote book reviews and editorial notes and saw the paper through to press.[107] All tasks that she would continue to fulfil in her new role.

The Listener had been established as a means of supporting broadcast talks, enhancing and contextualising the spoken word and promoting

---

[104] Rhoda Power Talks File 2, Somerville to Mallon, 16 March 1943.
[105] National Library of Scotland: Janet Adam Smith papers, ACC12342 (hereafter JAS papers), 122, Letter to Lambert, 7 November 1930.
[106] Oral History of the BBC: Janet Adam Smith interview, 1 August 1979.
[107] JAS papers, 122, letter to Lambert, 7 November 1930.

adult education. It turned out to be hugely controversial and, to avoid what was perceived as unfair competition, was severely restricted in the amount of non-BBC-related material it could publish.[108] A key challenge for Adam Smith was how to bring intellectual rigour within these constraints; however, the quality of the journal meant that the strict boundaries were gradually loosened. She was immensely influential and is credited with bringing the likes of W.H. Auden and Stephen Spender to a wider public. This was a time when modernist poetry was still viewed by many with suspicion and her four-page poetry supplement in 1933 (which included wood cuts by Gwen Raverat), resulted in many complaints.[109] Summoned by Reith to explain the outrage, Adam Smith succeeded in persuading him to allow T.S. Eliot to review the place of poetry within *The Listener*, a ploy that evidently calmed the situation.

Her handover notes to J.R. Ackerley who took over the Assistant Editor role in 1935 reveal that along with responsibility for poetry, she oversaw art articles and exhibitions, was the point of liaison for all book reviews, managed the published talks that were connected with books and art and selected short stories. This was in addition to a raft of general duties such as deputising for Lambert, attending editorial meetings and consulting on what was to be published in the journal. Her extensive correspondence shows in what high esteem she was held. She met her future husband, the poet and scholar Michael Roberts, through her work, although it was their shared love of mountain climbing that seems ultimately to have brought them together. Adam Smith had broadcast her first talk on mountaineering for BBC Scotland in 1932 and she continued to give radio talks, as well as to contribute articles to *The Listener*, after she left the BBC in 1935. On her departure to be married, Gladstone Murray, the Controller of Public Relations, wrote of his conflicting emotions; her loss was 'irreplaceable' but he 'rejoiced' in her happiness.[110]

Janet Adam Smith's initial letter to Lambert, requesting that she be considered for the Assistant Editor post, had made it adamantly clear she did not consider her work as an 'interesting way of passing the time

[108] See Asa Briggs (1965) *The Golden Age of Wireless: The History of Broadcasting in the United Kingdom*, Vol. 2 (London: Oxford University Press) pp. 280–92; Lambert, *Ariel and All his Quality*, pp. 91–108.

[109] *The Listener*, 21 January 1965, Janet Adam Smith, 'T S Eliot and The Listener'.

[110] JAS papers, 183, Gladstone Murray to Janet Adam Smith, 17 January 1935.

between going down from university and getting married'. She was 'quite certain' that to get the most of a job it should be regarded as part of a life-time career.[111] There is no doubt that if she had wanted to stay at the BBC she would have sailed through the Marriage Tribunal. But her six years on *The Listener* were without question of huge import to the journal and she left a legacy that was remembered for decades to come.

Social Documentary Maker

In 1937, Olive Shapley was promoted from *Children's Hour* to a new role in Manchester as a documentary feature maker. Archie Harding, the Director of Programmes for the North Region, had begun experimenting with new programme techniques and styles, working with the producer Geoffrey Bridson.[112] Shapley joined this small team where she would become the queen of 'actuality' broadcasting. The way in which she created her programmes, using a mobile recording unit, is vividly described in her autobiography. First used by Bridson, this was a large lorry packed with technical equipment which, for the first time, made it possible to record on location. It was Shapley, however, who realised its full potential, using it to bring to the microphone men and women who would never have entered a BBC studio. The scope of Shapley's work would be remarkable. Scannell and Cardiff, in their history of broadcasting in the interwar years, devote several pages to her pioneering work.[113] Through her novel recording technique, she made trail-blazing programmes such as *£.s.d.* where she visited local shops in the Calder Valley to learn how people spent their money. For *Night Journey* she delved into the lives of the long-distance lorry driver, travelling for hours in unsavoury conditions and even spending the night in a ditch. When the engineer who was with her learned that her next assignment would be down a mine his response was, 'with any luck, I shall be on holiday'.[114]

Other documentaries she made included *Canal Journey* (recorded on the Leeds–Liverpool canal), *Hotel Splendid* (a snapshot of a luxury hotel in Scarborough) and *The First Five Years*. This was a dramatic interpre-

---

[111] JAS papers, 122, letter to Lambert, 7 November 1930.
[112] Scannell and Cardiff, *A Social History of British Broadcasting*, pp. 334–5.
[113] Scannell and Cardiff, pp. 344–9.
[114] Shapley, *Broadcasting: A Life*, p. 52.

tation of nursery schools with recordings made at the Rachel McMillan Nursery School, where Shapley had originally trained. Amongst the cast were Joan Littlewood and James Miller, who later gained fame as Ewan MacColl.[115] In September 1938, under the headline 'Producer in Search of a Programme' *Radio Times* included a full-page article by Olive Shapley in which she described her adventures gathering material for *Homeless People*.[116] Traversing the north of England she had visited a Hospital for Incurables and Cripples run by French monks in the Yorkshire Dales, met abandoned youngsters in a Children's Garden Village near Manchester and encountered a school for young tramps near Durham. The evidence of distress, of grinding poverty had never seemed so apparent, she declared. Later, she would describe how the programme had been a 'rude awakening' for her, and for listeners.[117]

In March 1939 Shapley broke new ground with her documentary *Miners' Wives*. After recording interviews at the Cragshead colliery in County Durham, she had the inspiration to take Mrs Emerson, the wife of a checkweighman, to meet her counterpart at Marles-Les-Mines, a mining village near Bethune. Shapley and Emerson stayed with a mining family for a week. When the programme was aired, Mrs Emerson was in the studio for the second half, to recount her impressions of how the two communities compared.[118] Shapley's seminal production work, broadcast in 1939, was *The Classic Soil*. Conceived and written by Joan Littlewood this was an update on Engels' *The Condition of the Working Class in England in 1844*. It was described in *Radio Times* as 'comparing the life of the Manchester working man and his wife today with the life lived by their ancestors a hundred years ago'.[119] Looking back at the programme 60 years later, Shapley reflected that it was 'probably the most unfair and biased programme ever put out by the BBC' but this indicates the autonomy within which she worked.[120]

Shapley left the permanent staff of the BBC shortly before war broke out in 1939. She had married a colleague, John Salt, and found she was unable

---

[115] *Radio Times*. 6 February 1938. Along with Wilfred Pickles, Littlewood and Miller regularly presented Shapley's programmes.

[116] *Radio Times*, 2 September 1938.

[117] Shapley, *Broadcasting: A Life*, p. 52.

[118] Shapley, pp. 53–4.

[119] *Radio Times*, 23 June 1939. Also in 1939, Shapley made *They Speak for Themselves* about the work of Mass Observation.

[120] Shapley, *Broadcasting: A Life*, p. 54.

to continue in an established role because of the marriage bar rules. It is not quite clear why she did resign, as she should have been eligible for retention under the new marriage bar rules.[121] There is no indication that she was resentful about this as she continued on a contract basis. Neither is there any sign that during her BBC career she had felt hard done by in terms of promotion or pay. In fact Shapley faced 'hidden' discrimination, as did many other of the Corporation's salaried female staff.

5.2 'ON THE SAME FOOTING AS MEN?' RECRUITMENT, MOBILITY AND PAY

On the surface, there appeared to be no ostensible difference in the careers of salaried women and men. However, unquestionably there were divergences, not least in attitudes and motivations towards the job. As Mary Agnes Hamilton observed wryly in *Our Freedom and its Results* 'a woman has got in some way to be rather better than the comparable male to get herself regarded as his equal'.[122] And certainly at the BBC, the work ethic among women was often demonstrably stronger than that of their male colleagues.[123] Hilda Matheson, for instance, wrote of her Assistants, Lionel Fielden and Joseph Ackerley, as 'my leisurely young men lounging at their desks', lacklustre about work that did not interest them.[124] The personal files of the BBC's salaried women reveal, instead, a very different characteristic, a tendency to overwork.[125] Val Gielgud, reflecting on his pre-war years with the Corporation, described a host of hopeless young BBC men who assumed they could make it any-where, anyhow; something the women were not part of, he declared.[126]

---

[121] During the Second World War, Shapley went to the US with her husband and developed the programme *Fortnightly Newsletter*.

[122] Mary Agnes Hamilton, 'Changes in Social Life' in Ray Strachey, ed. (1936) *Our Freedom and its Results by Five Women* (London: Hogarth Press) p. 264.

[123] This necessity to constantly prove oneself was true of many professional women at this time. See, for example, Oram, *Women Teachers*, p. 84. Alix Meynell (1984) *Public Servant, Private Women: An Autobiography* (London: Victor Gollanz) p. 86.

[124] Hilda Matheson Letters, 1 May 1929. Matheson described Fielden as being 'very naughty about poetry readings because he is bored with them and forgets to see about copyright and things', 15 January 1929.

[125] See, for example, L1/1698/1: Beatrice Hart Staff File, Broadbent to Wade, 9 September 1933. In Miss Hart's case, overwork led to illness.

[126] Val Gielgud, History of the BBC: BBC Memories, available at: http://www.bbc.co.uk/historyofthebbc/bbc-memories/val-gielgud (accessed 6 May 2015).

This public school attitude meant that men often squandered opportunities which women would have grabbed.[127] Noel Annan, in his depiction of the ubiquitous 'old boys' network' in Britain in the 1920s and 30s, was keenly aware that women were disadvantaged. They found 'few ladders to help them climb'; they had 'few helping hands. They made it by talent alone', although he added the proviso 'unless they had beauty as well as brains'.[128] That there were many talented women at the BBC is manifest. But sexual inequality was deeply ingrained and in terms of recruitment, promotion and pay women were often treated very differently from men.

Inequality in Recruitment

In her 1934 book *Women and a Changing Civilisation* Winifred Holtby devoted nine pages to what she described as 'The Inferiority Complex', a woman's doubt in her own capacities.[129] This is one explanation as to why so many salaried women began their BBC careers as waged members of staff. More than half of those who held salaried posts had started this way, which compares starkly with the 90 per cent of men who were appointed directly to the salaried grades.[130] Unquestionably many young men, especially those who came from privileged backgrounds, exuded an air of entitlement to the influential jobs they quickly held at the BBC. The memoirs of Val Gielgud, Eric Maschwitz and Lance Sieveking make this immediately apparent.[131] The three were recruited on far bigger salaries than the £260 starting rate which, as will become clear, was a major reason for unequal pay.[132] While scores of men were recruited

[127] D.L. LeMahieu (1988) *A Culture for Democracy: Mass Communications and the Cultivated Mind in Britain between the Wars* (Oxford: Oxford University Press) pp. 10–11.

[128] Noel Annan (1985) *Our Age: The Generation that Made Post-War Britain* (London: Harper Collins) p. 12.

[129] Winifred Holtby (1934) *Women and a Changing Civilisation* (London: Lane and Bodley Head) pp. 97–105. Vera Brittain also eloquently described this in her article 'The Whole Duty of Women, *Time and Tide*, 23 February 1928, in Paul Berry and Alan Bishop, eds. (1985) *Testament of a Generation: The Journalism of Vera Brittain and Winifred Holtby* (London: Virago) pp. 120–3.

[130] Salary Information Files. Of the 702 men, only 60 had joined as weekly paid staff.

[131] Val Gielgud (1947) *Years of the Locust* (London: Nicholson and Watson); Eric Maschwitz (1957) *No Chip on my Shoulder* (London: Herbert Jenkins); BBC/WAC:S61: Special Collections: Autobiographical Sketches of Lance Sieveking.

[132] Gielgud on £450, Maschwitz on £325, Sieveking on £600. Sieveking and Maschwitz joined the BBC in 1926, Gielgud in 1928.

directly to high grade jobs, only three women arrived at the BBC on £500 or above before the Second World War, Hilda Matheson, Mary Adams and Christine Orr.

The 'old boys' network was a key element of inequality in recruitment; a recurrent question at interviews was which public school had been attended.[133] The rapid growth of the BBC generated an almost insatiable demand for programme makers, administrators, publicity staff, technicians and so on. Without formal entry requirements, at least until the mid-1930s, the quickest and easiest way to fill these roles was through personal contacts and on the whole, men suggested men. Being oblivious to women was characteristic of the upper echelons of management in the interwar years and similarly at the BBC, women were simply not in the frame. Although they might be considered if they put themselves forward or were suggested for BBC employment, the person who came most readily to mind was male.[134] As Mary Agnes Hamilton pointed out, it did not occur to men when contemplating such issues 'that they were doing anything out of the way'.[135] Reith was adamant that it was the 'duty' of the BBC 'to get the best man however we may come upon him'.[136] Val Gielgud's arrival is a case in point. Looking back on his long career from the vantage point of 1947, he declared his support for 'nepotism and favouritism' believing, correctly, that his own appointment in 1928 was due to the insistence of his two close friends Lance Sieveking (then a Talks Assistant) and Eric Maschwitz (then Editor of *Radio Times*).[137] Sieveking jested that in 1926 he was getting 'bright young men on to our staff at the rate of about one a week'.[138]

The introduction of Appointment Boards in 1934 did not markedly improve the situation for women; four-fifths of Boards interviewed only

[133] LeMahieu, *A Culture for Democracy*, p. 183. For a discussion on the 'old boys' network see, for example, Krista Cowman and Louise Jackson eds. (2005) *Women and Work Culture in Britain c. 1850–1950* (Aldershot: Ashgate) p. 15. This is much of the premise of Noel Annan, *Our Age*, pp. 26–68.

[134] Virginia Woolf noted men's propensity to appoint men. Woolf, *Three Guineas*, pp. 217–31.

[135] Strachey, ed., *Our Freedom*, p. 257.

[136] BBC/WAC:R13/216/1: Schools Broadcasting Department, Reith to Carpendale, 25 May 1935.

[137] Val Gielgud, *Years of the Locust*, p. 46. 'Lance Sieveking urged my intelligence upon Roger Eckersley, at that time Director of Programmes. Eric Maschwitz murmured of my merits into the ear of Gladstone Murray, Director of Public Relations'.

[138] Sieveking, *Autobiographical Sketches*, p. 50.

men.[139] In fact the BBC's increased bureaucratisation and the march towards professionalism appear to have made it harder rather than easier for women to get salaried positions. In 1939, a junior staff member, Daphne Parsons, drew attention to the fact that some advertised jobs were not open to women. These were identified as Editor and Deputy Editor of *The Listener*; the European Public Relations Officer; the Assistant Director of Office Administration; the Empire Programme Organiser; the Glasgow Director, the Outside Broadcast Manager and the position of Television Studio Manager.[140] The list is all the more surprising as at least one of these positions had, in the past, been held by a woman. It was conceded that, while there were occasionally vacancies for which women might not be wanted, no female member of staff was in fact debarred from applying.[141]

Inequality in Promotion

As early as 1926, Reith had decreed that Women Assistants should rank on the same footing as men and be as eligible as men for promotion.[142] But while women did rise through the BBC ranks, they were generally not promoted as quickly or as highly as men. Unlike teaching or the Civil Service, the BBC had no set paths for promotion. Rather, longevity of service, increased experience and good performance saw an individual rise through their grade until they reached the 'roof', at which point they could be considered for promotion to the next. This meant that subjectivity rather than a defined principle was at the heart of the process. In the same vein, while men were often fast-tracked by management to positions of authority, this was exceedingly rare for women, Hilda Matheson the only example. The fact that men were also geographically more mobile than women was also advantageous to their careers. For example, the position of Station Director, which was never held by a woman, was seen as a stepping stone to management responsibility at Head Office.[143] There was awareness of the limitations in opportunities for women's advancement,

---

[139] BBC/WAC:R49/27/1-3: Appointment Boards.

[140] This list was of all vacancies between 6 February and 12 April for which women had not been asked to apply. BBC/WAC:R49/739: Staff Vacancies, Pym to Clarke, 21 April 1939.

[141] Pym to Nicolls, 28 April 1939.

[142] BBC/WAC:R49/940: Women Assistants, 1926, Reith to All Station Directors, 30 April 1926.

[143] For example, Nicolls and Clarke had been Station Directors.

for instance, in 1929 it was specifically noted at Control Board that there was a tendency to consider men over women for promotion.[144] However, this did not preclude a Control Board agreement in 1932 'to get a half a dozen first-class men who would be posted in various changing places in Head Office and the provinces to learn the business'.[145] The creation of high-flying career opportunities for women was never considered.

Yet it appears that BBC women rarely sought out opportunities for advancement or looked to move into other areas of work. Winifred Holtby's 'inferiority complex' theory may have been at play here with women content to remain in jobs they knew well and which they were good at. Annual reports only occasionally indicate frustration with their roles. There was also the issue of loyalty. For instance, Margaret Hope Simpson, a graduate of Newnham, joined the BBC in 1931. She was briefly Hilda Matheson's secretary but from 1932 until 1943, she was secretary to Cecil Graves, staying with him as he progressed from Empire Services Director to Controller (Programmes) to Deputy Director General and ultimately Director General. Graves was aware that his reluctance to lose Hope Simpson was potentially holding back her career. It was only when he retired on ill-health in 1943 that she eventually moved on (she retired in 1964, an executive in Overseas Programme Planning). Mary Agnes Hamilton, writing specifically about the BBC, noted of the private secretary that 'the more useful she is, the less does her chief want to part with her'.[146] The one position that appears to have acted as a stepping-stone was that of Children's Hour Organiser, with a number of women moving into more senior posts. The majority of the BBC's salaried women, though, had linear careers and remained broadly in the same department and often in the same role for the entirety of their career with the Corporation. This had a limiting effect on how far within the BBC they could go.[147]

Inequality in Pay

Equal pay was a highly contentious issue in the interwar years. It was a focus of feminist campaigners and professional women's trade unions,

---

[144] BBC/WAC:R3/3/5: Control Board Minutes, 26 March 1929.

[145] Control Board Minutes, 15 November 1932.

[146] Mary Agnes Hamilton (1944) *Remembering my Good Friends* (London: Jonathan Cape) p. 283.

[147] Of the 128 women included in Salary Information only 19 had worked in more than one area of work during the interwar years and only a handful in more than two.

with teachers and Civil Servants at the forefront of the struggle.[148] The
BBC, unlike teaching and the Civil Service, did not operate separate salary
scales for women and men. All salaried staff were graded from 'E' (lowest)
to 'A' (highest). However, the process of recruitment, coupled with the
realities of promotion, meant there were proportionally far more women
in the lower grades 'E' and 'D' (56 per cent of women as compared to 25
per cent of men) than the top grades 'B' and 'A' (11.5 per cent of women
compared to 39 per cent of men). Only two women, Hilda Matheson and
Mary Somerville, were ever graded 'A', as Chapter 6 will show.[149] And
being in a lower grade was, naturally, associated with lower pay.

A chief reason for discrepancies in earnings at the BBC was starting
salaries. Almost invariably, those promoted from the waged ranks started
their salaried careers on £260 a year. Similarly, women recruited from
outside the BBC were likely to start on this rate, often thrilled with the
amount. Olive Shapley, for instance, recalled that as she sat in the studio
on her first day watching a BBC colleague play a tune on a tin whistle,
she thought 'they can't be going to pay me £250 a year for doing this'
(in fact, as noted, her starting pay was £260).[150] In *Our Freedom and
its Results*, Ray Strachey commented on how the task itself appeared to
absorb able women who were not active in demanding increases of pay
and status. It was an observable fact, she noted, that women were apt to
be 'tame, timid and biddable employees'.[151] At the BBC, all salaries were
independently negotiated which resulted in huge variations between staff,
men and women alike. For women, this could have a doubly detrimental
effect because not only were they less adept at negotiating higher starting
rates (if they were even aware that salary levels could be discussed) but
also they were far more likely to have come to the BBC from a lower paid

[148] See, for example, Winifred Holtby, 'Fear and the Woman who Earns', *News Chronicle*,
9 March 1934; Vera Brittain, 'Women still wait for Equality', *Daily Herald*, 26 March 1938.
Both quoted in Berry and Bishop, *Testament of a Generation*, pp. 81–3; 144–6. Strachey,
Careers and Openings, pp. 69–76. In the Clerical/ Executive grades of the Civil Service
although starting salaries were the same, the rate of increase for women was less and they
were ultimately paid three-quarters or four-fifths the salary of their male colleagues. The
Burnham Salary Scales for teachers specified that women should earn 80 per cent that of
men. Glew, 'Women in the GPO', p. 89, Oram, *Women Teachers*, p. 25.

[149] For those in Grade C, the proportion was more equal, 35.5 per cent of men compared
to 32 per cent women. These figures are derived from two sources: Salary Information Files
and BBC/WAC:R49/231/1.

[150] Shapley, *Broadcasting: A Life*, p. 34.

[151] Strachey, *Our Freedom*, p. 164.

job. This was recognised in 1938 when it was pointed out that although 'theoretically' the same payment was made to women and men for comparable work, because starting salaries related to outside market value, women generally began lower in the salary scales so 'age for age they were generally paid less'.[152] Reith was also attune to the significance of starting salaries. Janet Adam Smith was startled to discover that she had been given a £100 salary rise just two weeks before she left the BBC. It had been specifically requested by Reith who was aware that, if Adam Smith were to work again, her future employer would ask what she had last earned at the BBC 'and it would be better to say £650 than £550'.[153]

On rare occasions women did negotiate high starting salaries. Mary Somerville and Mary Adams, for example, and also Christine Orr. Orr refused to take less than £500 a year as Edinburgh Children's Hour Organiser (she had been earning £600 as a freelance journalist) even though this was almost double the usual rate for the role.[154] Annual increments were also linked to the grade, with those on £500 a year or above regularly receiving a pay rise of £50 or more.[155] Even in the lower grades, while for women a £20 annual rise was the norm, for men it was frequently more which suggests that they were often valued more highly and also that they were better at securing a higher rate.

A few women, like Doris Arnold, successfully bargained for above average salary rises.[156] Others, like Mary Hope Allen had an arduous battle to secure her £500 a year. Allen, the only woman in the BBC's experimental Research Section earned, in April 1930, £280 a year. Her three colleagues Lance Sieveking, Archie Harding and E.J. King Bull earned respectively £800, £500 and £400. In March 1931 she wrote to Reith complaining that her salary had only accumulated 'dribble by dribble' and stating that her financial ambition was 'to earn as much as the smallest salaried man in my section'.[157] After much management deliberation, in which it was emphatically denied that women were treated differently from men, she was granted a salary rise of £100. Olive Shapley is another example of an undervalued woman. It was recognised in April 1939 that she earned

---

[152] BBC/WAC:R49/605: Standardisation of Salaries, Pym to Nicolls, 11 November 1938.
[153] Janet Adam Smith interview.
[154] BBC/WAC:L1/328/1: Christine Orr Staff File, Orr to Dinwiddie, 7 October 1936; Orr to Pym, 13 November 1936.
[155] Salary Information Files.
[156] DAF, Beadle to Clarke, 11 October 1933.
[157] MHAF: 2, Undated letter from Allen to Reith, c. March 1931.

substantially less than her colleague Geoffrey Bridson (£425 to his £700) even though both were regarded as exceptional. Bridson was paid more because his market value was higher—there were fears that he might leave the BBC for film work.[158] In response, Shapley was regraded and her salary was increased to £600.[159]

There is no indication that Shapley ever felt dissatisfied with her pay (she had never pushed for the rise), and this acceptance of women's lower earnings, both by the individual and their managers, is a further reason why BBC women often earned less than men. This was acknowledged in the BBC's submission to the Royal Commission on Equal Pay in 1945. In preparation for the report, the Director of Staff Administration, Pym, informed William Haley, the new Director General, that although there were no posts in the Corporation in which different rates of pay were assigned to men and women for equal work, it did occasionally happen a woman was, 'put into the lower grade rather than the higher, or take longer to obtain her promotion, simply because she is a woman and not because her particular qualifications or performance in the post are below that of a man'.[160] Managers, when prompted, might rectify unequal pay but it frequently continued unobserved.

The Case of Mary Candler

An example of the ambivalent treatment of women is provided by Mary Candler. Candler joined the Copyright Section as a shorthand typist in 1928 and quickly proved herself irreplaceable taking on duties far beyond the remit of the job. Her Confidential Report for 1931 included a request from her manager, Dick Howgill, that she be regraded as she was 'definitely in the Assistant category'.[161] Shortly afterwards, there was a discussion about who should be Number Two in the Copyright Section, Miss Candler or a man. As neither had specific musical knowledge, Candler's greater experience won her the job, although she maintained her £5 weekly wage. It was made quite clear though that, should a suitably qualified man be found, he would be placed above her. In the meantime,

---

[158] BBC/WAC:L1/1821/1: Geoffrey Bridson Staff File, Confidential Report 1938.

[159] BBC/WAC:L1/1,783/1: Olive Shapley Staff File, Pym to Nicolls, 8 June 1939.

[160] BBC/WAC:R49/177: Staff Policy: Equal Pay for Men and Women, 1943–46, Pym to Haley, 16 May 1944.

[161] BBC/WAC:L1/799/1, Mary Candler Staff File 1 (hereafter MCF), Confidential Report 1931.

however, 'Miss Candler should get her chance'.[162] In April 1933, having performed the job well, Candler was promoted to the salaried grades on £280 a year.

However, in June 1933, Hamilton Marr, a barrister with musical expertise was appointed to the Section. Although Marr was to be Candler's senior, he had no knowledge of copyright, a subject she was expected to teach him. Candler was justifiably perturbed by this, especially as she imagined him to be earning £400 (she was correct) whereas the roof of her Grade, 'E' was £300. In a meeting with the Establishment Officer, Douglas Clarke, in which her attitude was described as 'in general excellent and entirely reasonable' she requested that her maximum salary should be raised to be at least equal to his starting salary.[163] The following year her salary rose from £280 to £300, the roof for her grade, which prompted Candler again to request regrading.[164] In this she was supported by her manager, Howgill, who confirmed that her qualifications and expertise were greater than those of Marr.[165] Yet, although Candler was recognised as 'the main spring of the section', she was not regraded 'D' until 1936, rising to Grade 'C' (with a salary roof of £600) in 1937, the same grade as Marr.[166] When Marr left the Corporation in 1939, Candler's obvious superiority saw her assume the position of Head of Copyright Section the following year. In 1942 she rose to be Copyright Director and in 1948 became the BBC's Head of Copyright, a position she held until her retirement in 1959.

Despite Candler's justifiable frustration at her treatment, there is no indication that she discussed the issue with female colleagues. Salaries and grading were a personal matter at the BBC. As far as we know, no salaried woman at the BBC was a member of a trade union, and even if there had been a staff association, it is unlikely these irritations would have been shared. Unlike their counterparts in teaching and the Civil Service, the Corporation's salaried women never felt sufficiently angry to protest, especially as sexual inequality was largely private and so unspoken. BBC women did not identify themselves as female workers with a shared grievance, rather they identified themselves with BBC men, as public servants

---

[162] MCF, Confidential Report 1931.
[163] MCF, Clarke to Nicolls, 1 August 1933.
[164] MCF, Candler to Howgill, 21 March 1934.
[165] MCF, Howgill to Nicolls, 6 April 1934.
[166] MCF, Confidential Reports, 1936, 1937. In April 1939, Candler earned £450, Marr £580.

involved with the crucial job of broadcasting. In their work they viewed themselves as equal even though, behind the scenes, many were not.

5.3   CONCLUSION

The BBC's salaried women worked in a wide variety of positions that were broadly viewed to be equal to those of men. They were bright, motivated women in the vanguard of Britain's post-suffrage generation, an era when new professional opportunities were opening up. Ray Strachey, for instance, celebrated the 'queer' jobs that were now available and which represented the adventurousness of women, the diversity of which was evident at the BBC.[167] The rapid expansion of the Company/Corporation in the interwar years meant that there was the chance to grasp the initiative and to develop new areas of work be it as a Drama Producer, a Librarian or an Advertising Representative. Personal files reveal women to be conscientious employees, praised for their hard work, loyalty and creativity. The BBC was constantly in the public eye and the employment of women was applauded by the press, their prominence in unusual areas of work and in well-paid roles played to the image of the BBC as a progressive, forward-looking organisation.

However, while there was an outward impression of modernity and equality, in reality BBC women faced unspoken discrimination in recruitment, promotion and pay. Salaried posts were frequently filled by personal recommendation and whether these were university friends, the relatives of acquaintances or nominated outsiders, it gave an advantage to men. Men also benefited from having more dynamic BBC careers. They were encouraged to move between departments; each new position offering not only a salary rise but wider experience, indispensable to those aspiring to reach the top. Salaried women, on the other hand, often remained in the same job for their entire career, gradually working their way up. They rarely put themselves forward for promotions nor were they earmarked by their managers for significant new posts. Most had started their BBC careers on the bottom rung of the salary ladder with significant numbers promoted from the waged ranks. This meant they were subsequently disadvantaged by lower annual increments while only a handful appear to have been as adept as men at negotiating higher rates. Nevertheless, by interwar standards, the pay and conditions of service at the BBC were

[167] Strachey, *Our Freedom*, p. 141.

good and most women accepted these without complaint. The majority earned more than £300 a year, a salary identified by Margaret Cole, as 'doing very well indeed'.[168]

Salaried women made a notable contribution to the BBC. They created acclaimed programmes, supplied visual images, oversaw copyright and supervised essential services. The BBC's unique position as the first broadcasting industry, with the novelty of creating what would become a British institution from scratch, gave unique opportunities for women. They were able to access challenging and fulfilling jobs and to progress to positions of seniority which influenced the development of the Corporation and contributed to its cultural growth. Four BBC women commanded exceptional salaries for the times and the next chapter considers the careers of Mary Somerville, Hilda Matheson, Isa Benzie and Mary Adams.

---

[168] Margaret Cole (1938) *Marriage: Past and Present* (London: Dent and Son) p. 153.

CHAPTER 6

'Women Who Rule at the BBC': Four Elite Women

In September 1959, the former BBC governor, Mary Agnes Hamilton, wrote an acerbic letter to *The Times*. Where were the women in top BBC jobs, she queried? 'Since the days of Hilda Matheson, Mary Somerville, Mary Adams and Isa Benzie', Hamilton claimed, 'women have not appeared in director posts'.[1] Hamilton's observation was pertinent. By the late 1950s there were proportionally far fewer BBC women in top jobs. The interwar BBC was in many ways a 'golden age' for high-fliers, particularly in the mid-1920s and early-1930s. This chapter focuses on the interwar careers of Matheson, Somerville, Benzie and Adams, four women who, on salaries of £900 or more, were the highest paid. Three of them were Directors. Hilda Matheson was Director of Talks 1927–1932, a time when the broadcasting of the spoken word came to maturity and was at its most controversial. Mary Somerville assumed the title Director of School Broadcasting in 1931, a service she had virtually created and would continue to lead until 1947. Isa Benzie, a former BBC secretary, was Foreign Director 1933–1938, becoming the public face of the BBC in its international relations. Mary Adams, the first woman Television Producer in 1937, did not attain a director post in the interwar years (she applied for but failed to become Director of Talks in 1936). However, post war, she would rise to be Assistant to the Controller, Television.

---

[1] Quoted in Michael Fogarty, A.J. Allen, Isobel Allen, Patricia Walters (1971) *Women in Top Jobs: Four Studies in Achievement* (London: Allen and Unwin) p. 166.

© The Editor(s) (if applicable) and The Author(s) 2016
K. Murphy, *Behind the Wireless*,
DOI 10.1057/978-1-137-49173-2_6

Already by 1937, the *Manchester Guardian* was posing the question 'Does the BBC want women chiefs?'[2] Although its subsequent claim that in the early days of the BBC there had been almost as many women heads of departments as men was an exaggeration, it did pinpoint a reality that by the mid-1930s, only men were selected for executive posts. In its first ten years or so, the BBC did promote women to high status roles; so what were the conditions that enabled this? It was not without precedent for professional women to hold high-salaried positions in the 1920s and 1930s. Head teachers of the largest girls' schools, university professors and senior Civil Servants could all earn salaries that neared £1,000.[3] As Matheson crowed to Vita Sackville-West in 1929:

> My BBC pays me a fat screw, £900 a year, which is more than almost any other woman I know gets and quite out of the way generous. I mean as women's pay goes it's more than many women get in responsible civil service jobs I know.[4]

Others were paid far more. At the John Lewis Partnership, three women earned in excess of £1,200 per annum while women directors of advertising companies, such as Florence Sangster and Ethel Wood, would have been on similar salaries or above.[5] Alice Head, the Managing Director and Editor of *Good Housekeeping* was reputedly the highest-paid woman in Britain, though it is hard to believe she was paid more than the City stockbroker, Beatrice 'Gordon' Holmes, who earned an eye-watering £4,000–£5,000 a year. Unsurprisingly, the press were fascinated by these high-flyers and pioneers.[6] Contributing to an article in *The Star* in November 1937, Ray Strachey declared 'Women in these big-salaried

[2] *Women in Top Jobs*, p. 165.

[3] Carol Dyhouse (1995) *No Distinction of Sex? Women in British Universities, 1870–1939* (London: UCL Press) p. 150; Dorothy Evans (1934) *Women and the Civil Service* (London: Pitman) pp. 151–8. Dorothy Evans identified two female senior civil servants who were on salaries in excess of £1,300 in 1934. The Director of Women Establishments earned an undisclosed salary.

[4] Hilda Matheson letters, (hereafter HML), 12 January 1929.

[5] *The Gazette* (the journal of the John Lewis Partnership), 23 January 1932. Sangster and Wood's salary is undisclosed but Biscoe estimated a Managing Director in advertising earned £1,000 a year. Vyrnwy Biscoe (1932) *300 Careers for Women* (London: Lovat Dickson) p. 18.

[6] Adrian Bingham (2004) *Gender, Modernity, and the Popular Press in Inter-War Britain* (Oxford: Clarendon Press) pp. 50–2, 63–8.

positions increase every year', though she added the caveat that 'women who win such appointments have to be better at their jobs than men holding corresponding positions'.[7]

Eileen Power, Professor of Economic History at the LSE (on a salary scale £1,000–£1,250), may have taken Strachey to task on this comment. She claimed never to have discerned any difference of treatment between herself and the men she worked with.[8] Power's biographer, Maxine Berg, pointed out, however, that she was in truth paid less than her male colleagues, was not put on any professorial appointment committees nor did she sit on any government enquiries or commissions.[9] Alix Kilroy, the first female civil servant to be promoted to the position of Principal in the Board of Trade, while stressing that her working relationships with male colleagues were ones of equality and respect, conceded that her promotion would have happened more quickly and 'been a foregone conclusion' for a man in her place.[10]

Yet there was undoubtedly a sense amongst high-powered women of this period, notably in the BBC and particularly in professions where they worked alongside men, that they were equals.[11] Harold Nicolson in his 1932 novel, *Public Places*, portrayed the character of Jane Campbell (based on Hilda Matheson) in this light.[12] The fictional Parliamentary Undersecretary of State for Foreign Affairs was 'a graduate of Lady Margaret Hall ... the modern woman, emancipated ... talking as man to man'.[13] The fact of being a pioneer meant that these women were often isolated and so 'novelties', but once the 'old boys' network had been breached, it appears that men were chivalrous and largely accepting of

---

[7] *The Star*, 19 November 1937.

[8] Maxine Berg (1996) *A Woman in History: Eileen Power 1889–1940* (Cambridge: Cambridge University Press) p. 181. This was her salary range in 1932.

[9] Berg, *A Woman in History*, p. 181.

[10] Alix Meynell (1984) *Public Servant, Private Women: An Autobiography* (London: Victor Gollanz) p. 129; Beatrice Gordon Holmes (1944) *In Love with Life: A Pioneer Career Woman's Story* (London: Hollis & Carter) pp. 82–4. Holmes was debarred from membership of the London Stock Exchange, which refused entry to women until 1973.

[11] For a discussion on the experiences and attributes of elite women see Fogarty, *Women in Top Jobs*, pp. 14–16, 34, 41. Kanter also investigated the circumstances of women in top jobs. Rosabeth Moss Kanter (1977) *Men and Women of the Corporation* (New York: Basic Books) pp. 206–29.

[12] Nigel Nicolson (2004) *Harold Nicolson: Diaries and Letters, 1907–1964* (London: Weidenfeld and Nicolson) p. 92.

[13] Harold Nicolson (1932) *Public Faces* (London: Penguin) p. 80.

their new female colleagues. Kilroy raised the apposite point that her self-confidence as an Oxbridge graduate meant that she took for granted the equality that her job at the Civil Service promised and that this may have influenced her reception and treatment.[14]

Kilroy had been to university yet reading the biographies and autobiographies of high-achieving women at this time one finds little similarity in their backgrounds or education.[15] Beatrice Gordon Holmes came from a lower-middle-class family, left school at 14 and started her working life as a £1 a week typist. Alice Head's first job also paid £1 a week, although she had the advantage of attending North London Collegiate. Both women, however, had the good fortune to be supported by male bosses who recognised their abilities and gave them the all-important leg-up. But whereas Holmes was evidently tenacious, Head (as suggested by the title of her memoir *It Could Never Have Happened*) was not particularly go-getting. Alix Kilroy grew up in an environment of women's suffrage and always expected to earn her living, the experience at Oxford opening her mind to the possibilities of a substantial profession.[16] Eileen Power, on the other hand, found her vocation while at Cambridge, a life-changing experience for a young woman whose childhood had been riven by scandal and tragedy.[17] The BBC's highest-paid women had similarly diverse backgrounds, so what was their journey into the top echelons of broadcasting?

6.1   Being Elite at the BBC

As might be expected, Mary Somerville, Hilda Matheson, Isa Benzie and Mary Adams were middle-class but none of their families were wealthy. Somerville and Matheson were both daughters of the manse, their fathers Presbyterian ministers. Benzie's father was a chartered accountant, Adams' a struggling farmer. It is notable that all except Adams were Scottish, which may have endeared them to Reith; in the case of Benzie, this was of great significance, as we shall see. None of the four women had been to public school, unlike the majority of the men they worked alongside. Somerville

[14] Meynell, *Public Servant*, p. 129.

[15] Virginia Nicholson provides a colourful overview of many successful women. Virginia Nicholson (2007) *Singled Out: How Two Million Women Survived without Men after the First World War* (London: Viking).

[16] Meynell, *Public Servant*, pp. 76–8.

[17] Berg, *A Woman in History*, p. 55. Power's father went to prison for fraud and her mother died when she was fourteen.

had attended Selkirk High School, although much of her education took place at home because of ill health (she was diabetic). Matheson was sent from her home in Putney to St Felix, a girls' boarding school in Southwold when she was 14. Benzie was educated at Laurel Bank School in Glasgow, as well as at private and convent establishments in Belfast and Southern Ireland. Adams, whose father died when she was 12, won a scholarship to Godolphin School in Salisbury.

All but Adams were Oxford graduates. Matheson's three years as an Oxford Home Student (where she studied History) reveal her to be a bright, cultured and physically active young woman, a member of the hockey team, the choir and the amateur dramatics society.[18] At Somerville College, Somerville (who read English Literature) held court at high-powered tea parties, where 'the conversation was far ranging', with Maisie, as she was called then, always at the centre.[19] Benzie, at Lady Margaret Hall (where she studied German), excelled at games, her popularity and gusto manifest in her selection as College Secretary, Head Student and President of the Junior Common Room. Adams, who gained a First in Natural Sciences at University College, Cardiff, then moved to Newnham College, Cambridge, as a research scholar, where she gained her MSc.

Matheson and Adams were both mature women when they came to the BBC in 1926 and 1930 respectively. Matheson's varied pre-BBC career included a post-university stint as part-time secretary to H.A.L. Fisher and war-time work that encompassed, from 1916, the Registry of the Special Intelligence Directorate (the precursor to MI5). After the war, she worked briefly for Philip Kerr (later Lord Lothian) before becoming, in 1919, Political Secretary to Nancy Astor, newly elected as MP for Plymouth (Sutton).[20] Adams, a biologist, had followed an academic career at Cambridge, during which time, in 1925, she married Vyvyan Adams who was elected Conservative MP for West Leeds in 1931. This was an intriguing match as Mary Adams was a pronounced socialist.[21] Somerville came to the BBC in 1925 directly from Oxford as did Benzie in 1927, albeit via a secretarial course at Pitman's College.

---

[18] Various Authors (1941) *Hilda Matheson* (Letchworth: The Hogarth Press), memories of Ruth Butler and Mrs H.A.L. Fisher, pp. 27–31, 39–42; Michael Carney (1999) *Stoker: The Biography of Hilda Matheson OBE, 1888–1940* (Llangynog: Michael Carney) pp. 6–9.

[19] Mary Somerville, 'A Tribute from her Friends', broadcast 31 May 1964, memory of Janet Vaughan.

[20] Biographical details from Carney, *Stoker*.

[21] *Oxford Dictionary of National Biography*. Mary Adams, entry 30750 by Sally Adams.

As non-Londoners, the four women set up home in the capital. Matheson shared a large house in Kensington with two friends. Here, she had her own bedroom and sitting room and the services of a housekeeper.[22] As befitted the wife of an MP, Adams lived in grandiose Gloucester Gate while Somerville, upon her marriage in 1928, moved to St John's Wood. Benzie, prior to her wedding in 1938, lived for many years in a Ladbroke Grove flat with her university friend Janet Quigley, who she introduced to the BBC. It was not unusual for unmarried professional women to live together in the interwar years. Eileen Power lived for a time with Karin Costelloe, Ray Strachey's sister; Vera Brittain and Winifred Holtby shared several flats while Alix Kilroy lived with her sister Mona, a Chief Buyer at John Lewis.[23] Although Matheson mused about buying her own flat, which she could easily afford, she enjoyed the company and emotional support of her house-mates.[24]

Adams and Somerville were consummate hostesses.[25] As wives, and later mothers, they had strong ties to the home which became their base for networking. This was extremely important for their work, much of which relied on building and maintaining contacts. Both women were granted the concession of entertaining at the BBC's expense; Somerville, for instance, claimed 7/6d per guest.[26] Their male counterparts at the BBC would all have been members of at least one Club, something that was encouraged and indeed paid for by the BBC.[27] Here they would have carried out what Holtby derided as 'oysters and champagne diplomacy'.[28]

---

[22] Matheson's 'Sumner Place Family' were Marjorie Maxse, the future Conservative MP and Dorothy Spencer.

[23] Berg, *A Woman in History,* p. 141. Vera Brittain (1940) *Testament of Friendship* (London: Fontana) pp. 113, 274; Meynell, *Public Servant,* pp. 105, 159. See also Nicholson, *Singled Out,* pp. 151–7.

[24] HML, 2 February 1929.

[25] For example, Matheson went to two parties at Somerville's new house, HML, 2 February 1929, 8 May 1929.

[26] BBC/WAC: Mary Somerville Staff File 2: (hereafter MSSF:2), Nicolls to Carpendale, 25 May 1934. There are extensive expense files for Mary Adams, see for example BBC/WAC:S322/39/2: Correspondence, Talks Department.

[27] Control Board Minutes often report the payment of subscription fees for Clubs. A few women profited from this. Dorothea Barcroft, Birmingham Children's Hour Organiser joined the Soroptimist Club in June 1928 and Kathleen Lines the Arts Theatre Club in April 1930. BBC/WAC:R3/3: Control Board Minutes, 26 June 1928, 10 April 1930.

[28] Winifred Holtby (1934) *Women and a Changing Civilisation* (London: Lane and Bodley Head) p. 90.

Matheson belonged to at least one club, the Albermarle (her membership funded by the BBC), but rarely used it.[29] She preferred instead to build and maintain her contacts through intimate lunches, dinners and at house parties. Her later membership of the Women's Provisional Club (founded in 1924 to serve high-flying business and professional women) lasted little longer than a year.[30] Benzie was also nominated as a member but her election was cancelled through lack of response.[31] Very possibly the intensity of her BBC job, as would have been the case with Matheson, militated against attendance.

Of the four women, only Adams requested a transfer to her new position in Television, all the others had 'fallen into' the role of Director. As closer scrutiny of their BBC careers will show, at management level their immediate contacts were with men and in many ways they were isolated within separate male spheres. Although there is some evidence of friendships, particular between Somerville and Matheson, there appears to have been little time for sharing concerns or socialising. The BBC was a demanding place to work, particularly for overtly conscientious women.

6.2   MARY SOMERVILLE (1897–1963)

At some point in 1927, Mary Somerville wrote a letter to Valentine Goldsmith, the BBC's Head of Administration.[32] Addressing him as 'Dear Val' she laid bare the dire financial situation she found herself in. The Kent Experiment into school broadcasting, she explained, had involved her driving 10,000 miles, necessitating a new car (for which she had taken out a BBC loan), and one that now needed a complete overhaul with 'new clutch, self-starter, re-varnishing etc.' As a result of the intense workload, she had also been unable to earn a 'ha'penny' extra, unlike the £95 she had added to her salary the previous year through writing. Her 'present screw' of £500 meant that she did not have enough to 'run a car, live suitably and keep out of debt'. A supplementary note cheekily berated the BBC for allowing 'a poor defenceless female work more than you let her earn'.

[29] HML, 12 December 1928.
[30] Women's Library:5/WPV/3/1:Women's Provisional Club (WPC): Executive Committee Minutes, 29 March 1931, 31 May 1932.
[31] WPC: Executive Committee Minutes, 14 September 1936, 25 January 1937.
[32] BBC/WAC:L2/195/1: Mary Somerville Staff File:1 (hereafter MSSF:1), 1925–35. Undated but between April 1927 and February 1928.

These private communiqués hint at the nature and character of Mary Somerville. Throughout her BBC career she would always be forthright in her views, she would always overwork, she would always maintain her links with the literary world and she would always be short of cash. The car was also important. Because of diabetes and other health problems, it was essential for her to drive. Her note to Goldsmith also reveals Somerville's situation and state of mind at a turning point in her BBC career. After two years in the Education Department, the gruelling Kent Experiment she fronted, as will be revealed, changed her perception of school broadcasting and cemented her position as its driving force.

It was while she was at Oxford, in April 1924, that Somerville recalled hearing her first radio broadcast. On a visit to a country schoolhouse, she shared two pairs of earphones with a schoolmistress and three pupils to listen 'by chance' to a BBC talk on music by Sir Walford Davies.[33] It was an early trial broadcast for schools and Somerville wrote of the profound effect it had on all five listeners. The impact was 'tremendous', they were 'exalted' and Somerville realised then 'what this brave new medium of communication might mean for schools'.[34] Within a week, she had arranged a meeting with John Reith who recorded in his diary a visit from 'a very clever and self-confident young lady', recommended by a mutual friend.[35] Reith later recalled how she had 'in effect told me that she was joining the BBC and the sooner the better'. He advised her to return to Oxford to finish her degree.[36] In February 1925, Somerville again wrote to Reith proposing that she join the BBC as an Assistant to the Director of Education, J.C. Stobart.[37] Stobart was keen on the idea. Somerville was expected to get a First; she had good contacts within the literary world; she had an attractive voice and the growth of school work meant she would be an asset to his department.[38] After some deliberations over start date and salary, Somerville joined the BBC in July 1925.

---

[33] Broadcast 4 April 1924.

[34] Mary Somerville, 'How School Broadcasting Grew Up' in Richard Palmer ed. (1947) *School Broadcasting in Britain* (London: BBC) p. 9.

[35] Reith Diaries, 11 April 1924.

[36] *The Times*, Mary Somerville Obituary, 6 September 1963.

[37] MSSF:1, Somerville to Reith, 24 February 1925.

[38] MSSF:1, Stobart to Carpendale, 4 March 1925. Somerville's literary connections included Robert Graves, Robert Bridges, Sir Edmund Gosse and George Moore, *Manchester Guardian*, 2 September 1963, *News Chronicle*, 16 January 1936. *News Chronicle* hinted that to enter broadcasting she gave up a promising literary career, her early short stories being 'acclaimed by discriminating critics'. Because of ill health Somerville was awarded an aegrotat degree.

When Somerville arrived at Savoy Hill, broadcasts to schools were rudimentary.[39] Since October 1924, five weekday talks were scheduled at 3.15 pm. These were half-hours on, for example, Dickens, British Plants or Music which were given by experts, such as Walford Davies, and by BBC staff. Somerville herself was soon broadcasting (hence Stobart's reference to her attractive voice), excelling in literary topics such as Modern Poetry, Shakespeare's Heroines and English Composition. As well as giving talks, Somerville was soon overseeing all transmissions to schools: suggesting subjects and lecturers; testing voices; advising new speakers on broadcast techniques; overseeing studios; planning teaching notes; conducting correspondence with teachers and educationists; visiting schools and preparing articles for educational papers.[40] However, the Education Section faced a very real problem. Few schools were listening hampered by the amateurishness of most wireless equipment and the style of the talks. There were also very real fears that, not only might broadcasting supplant teaching, but also that the BBC could potentially influence the educational agenda. In 1924, 220 schools were listening to school broadcasts; by 1927 this had risen to around 3,000.

The Kent Experiment changed all this.[41] Funding had been provided for good quality receiving equipment to be installed in a selection of schools in the county.[42] Somerville's eight-month secondment to the Experiment saw her visiting these schools, watching lessons and talking to teachers and pupils about their experience of school broadcasts. It quickly became clear that programmes were failing because producers and broadcasters were unaware of how children learnt: they used too many complex words; they spoke too fast; they patronised; in effect, they failed to captivate and inspire.[43] Commenting about the Kent Experiment in 1954, Somerville recalled that she had found 'poor, patient children sitting in rooms being bored ... It wasn't that the programmes were bad ... but they were not produced or given by people who

---

[39] For a more detailed history of school broadcasting see Asa Briggs (1965) *The Golden Age of Wireless: The History of Broadcasting in the United Kingdom*. Vol. 2 (London: Oxford University Press) pp. 185–218; Mary Somerville, 'How School Broadcasting Grew Up', pp. 9–16.

[40] BBC/WAC:R13/419/1/ Departmental: Talks Division: Talks Department 1923–29, Duties, Talks Section, October 1926.

[41] For specific details of the Kent Report see Briggs, *Golden Age of Wireless*, pp. 190–5.

[42] The Carnegie Trust funded the experiment. Kent was chosen partly because of its range of urban, rural and semi-rural schools.

[43] Somerville, 'How School Broadcasting Grew Up', pp. 12–13.

knew children in their bones'.[44] As Hilda Matheson also quickly realised, talks needed to be written specifically for broadcast. When it came to schools, careful scripting and editing was even more important to win over teachers and pupils to the new medium of radio.

Although positive about the value of school broadcasts the final Kent Report, published in 1928, indicated a number of areas for improvement, especially in the way programmes were delivered. In response, Somerville developed a range of new techniques and introduced new speakers. One of her discoveries was Rhoda Power who broadcast the first of her innovative history series that year (see Chapter 7).[45] Somerville also pioneered other new styles of broadcasting such as Ann Driver's *Music and Movement*, the first to be aimed at infants.[46] The physiologist, Professor Winifred Cullis (who would become a stalwart of school broadcasting) found herself at the receiving end of Somerville's determination to improve performance, as did Walford Davies. Prior to her first talk, Cullis was taken by Somerville to watch a school audience listen to a programme, 'the result of which ... she made haste to revise her own talk entirely, for she had forgotten how small one is at eleven'.[47] Walford Davies was provided with a 'watchdog', his description of the observer employed to comment on his delivery.[48]

A second development from the Kent Experiment was the expansion of support materials with 233,000 pamphlets issued in 1927, output overseen by Somerville.[49] A third was improved wireless reception in schools through specially trained BBC Education Engineers, a system also initially overseen by her.[50] The most significant development, however, was the inauguration of the Central Council for School Broadcasting (CCSB) with Somerville as Secretary. By the time the CCSB first met in February 1929, Somerville had become de facto Head of School Broadcasting. Stobart, whose health was poor, had increasingly taken a back seat and she was his natural successor.

---

[44] Rough Notes for Co-ordinating Committee Discussion, October 1954, as quoted in Briggs, *The Golden Age of Wireless*, p. 195.

[45] 'Boys and Girls of Other Days' was broadcast weekly from 17 January 1928.

[46] 'Music and Movement' was first broadcast 28 September 1934.

[47] *Good Housekeeping*, August 1935.

[48] Somerville, 'How School Broadcasting Grew Up', p. 12.

[49] Briggs, *The Golden Age of Wireless*, p. 195.

[50] BBC/WAC:R16/537: School Broadcasting Memos and Reports, Somerville to Local Education Authorities, 22 July 1930.

The CCSB secured links between the BBC and outside bodies such as the Board of Education, Local Education Authorities and individual teachers. Six programme sub-committees, most of whose members were teachers and specialists, helped draft the school broadcasting syllabus in their subject area for the coming year. They also advised on the commissioning of scripts, edited educational pamphlets and recommended potential speakers. A vital link between the Programme Sub-Committees and the BBC Schools Department was the BBC Education Assistant. The Kent Experiment made Somerville aware of the need to employ former teachers in this capacity and a specially recruited BBC Assistant sat as secretary to each sub-committee. This link ensured the BBC, under the direction of Somerville, could implement the Committee's recommendations. Always canny, Somerville negotiated the perk of extra leave for her Education Assistants during school holidays, in order to bring the Department more in line with school terms.[51] Richard Lambert was incredulous about this privilege, which was unknown in other departments.[52]

In 1925, one 30-minute talk had been broadcast daily during term-time. By 1929 this had been extended to 1 hour and, in 1935, 2 hours of schools programmes were generally available each day. As the number of school talks increased, so did Somerville's workload. A 'rough sketch' of her work in August 1933 included, amongst normal office routine: visiting schools; summarising the annual teachers' criticism of programmes; translating the recommendations of the Programme and Pamphlet Committee; reviewing the activities of the Education Engineers and drafting the reports of Programme Sub-Committees.[53] This was without a raft of extra duties such as attending conferences, meetings and official inquiries. Somerville also constantly battled for extra money for her department, for example there were attempts to persuade the Board of Education to fund wireless equipment in schools but with no success.[54] The pressures were immense and in September 1934, Somerville collapsed from stress, necessitating six months leave.

On her return it was decided that one way her workload could become more manageable was by restructuring the CCSB. As a result, it became a

---

[51] BBC/WAC:R13/216/2: School Broadcasting Department:2.
[52] Lambert (1940) *Ariel and All his Quality* (London: Gollanz) p. 54.
[53] BBC/WAC:R49/611/1: Schools Broadcasting, Somerville to Siepmann, 26 August 1933.
[54] Briggs, *The Golden Age of Wireless,* pp. 201, 206.

separate, autonomous public body, with a new Secretary, A.C. Cameron.[55] When Cameron took on his new duties in November 1935, his salary was £1,500, significantly more than the £1,100 Somerville had been paid for the dual role.[56] Somerville's BBC salary had long been a cause of tension, the £400 she had negotiated in 1925 viewed as 'a very reasonable salary for a girl of her years'.[57] Unimpressed, in every subsequent year she demanded a hefty rise and had achieved £600 by the time of her maternity leave in 1929, as Chapter 4 disclosed.[58] Following her return to the BBC after the birth of her son, Somerville's salary rose to £750 in January 1930.[59]

Initially content, Somerville then learned that her colleague Charles Siepmann, promoted to Head of Adult Education in late 1928, was earning far more than her.[60] At first managers justified the differentiation by defending the 'controversial and delicate nature of Adult Education talks'.[61] Later they conceded that 'we have shown some weakness in the matter in that we have not handed things out to her as to the men, but have waited for her to complain and prove her case'.[62] Somerville never expressly stated that inequalities in her salary were due to her gender, however, the discrepancy arose at the point she was negotiating maternity leave. Once the pay gap emerged, it was allowed to continue until Somerville demanded redress. Even then, it never completely closed and, with Siepmann's promotion to Talks Director in 1932, it widened considerably.[63]

Although she may have been privately disgruntled, Somerville relished her BBC job as her 'BBC Diary' in *Radio Pictorial* in October 1935 illuminates.[64] This particular day started with a call from her secretary, Miss Scott, reminding her to be at the office early to read through a memorandum

[55] Cameron was formerly Director of Education for Oxford.

[56] Somerville was shocked when she first became aware of how much larger Cameron's salary was than hers. Mary Somerville Papers, Letter to Briggs from Rowan Davies, 21 February 1965.

[57] MSSF:1, Eckersley to Goldsmith, 13 September 1926. Somerville's salary compared favourably to her male colleagues. In 1927, as Education Assistants, George Dixon earned £200, Derek McCulloch earned £350 and, arriving in 1928, Tony Rendall earned £280.

[58] MSSF:1, Confidential Report, February 1929.

[59] MSSF:1, Confidential Report, February 1930. Somerville was acknowledged as 'virtually Schools Director'.

[60] Siepmann joined the Adult Education Section in 1927 as an Assistant.

[61] MSSF:1, Graves to Eckersley, 6 March 1930.

[62] MSSF:1, undated, unsigned but most probably sometime in 1933.

[63] In 1933, Siepmann earned £1700 to Somerville's £950.

[64] *Radio Pictorial*, 4 October 1935.

on next year's programme commitments. Then followed a whirl of meetings and telephone conversations peppered with broadcast listening, script reading and a visit to an infant school in the East End. Lunch and dinner (presumably at the BBC's expense) involved hosting academics which meant talking 'shop' from 8 am until 11 pm. Somerville's marriage to Ralph Penton Brown, the *Morning Post* correspondent in Belgrade, had not been a success and her 'BBC Diary' offered a glimpse into her home life, where she lived apart from her husband.[65] Mrs Bishop, Somerville's 'beloved, long-enduring, undefeatable' housekeeper, ran the house and helped care for six-year-old Timothy whom Somerville described, tongue-in-cheek, as 'the "neglected child" of a working mother'. Mrs Bishop also prepared meals although Somerville was quick to point out that she liked to make the sauces. She also spent as much time as she could with Timothy, on this occasion arriving home by 6.30 pm to play Meccano with him and to bathe him. The diary is an ebullient mix of frantic BBC work and pleasurable domesticity. Somerville was rare in being a working woman with a young child who was at the peak of her game.[66] The *Radio Pictorial* diary presented her both as a top-class executive and a devoted mother and Somerville was at pains to make clear that, although it was a struggle, there was no tension between the two.

By 1937, Somerville's salaried team at Head Office amounted to ten: six men and four women. Between them they oversaw 27 different educational courses which included tailored talks for infants, juniors and seniors, all of which were carefully prepared with extensive support materials. By 1939, the number of schools listening to the broadcasts had reached almost 10,000. Throughout the late 1930s, Somerville continued to demand better resources and in June 1939, the staffing situation became so dire that she threatened to resign.[67] Before the situation had been fully resolved, however, the war had started which involved Somerville and the Schools Department in major shifts, both of location and programmes.[68]

---

[65] A memo states that Somerville's husband lived in the Balkans and did not contribute to the household expenses. MSSF:1, Nicolls to Carpendale, 30 July 1934.

[66] It is hard to think of many examples. Sarah Lewis, the wife of John Spedan Lewis, continued to work for the John Lewis Partnership, as a Director of Peter Jones and Deputy Director of the Partnership, while raising three young children. Other senior professional women who were mothers and who worked included the haematologist Janet Vaughan and the crystallographer, Kathleen Lonsdale.

[67] MSSF:2, Clarke to Pym, with handwritten additions by Rose Troup, 7 June 1939.

[68] Asa Briggs (1979) *The War of Words, The History of Broadcasting in the United Kingdom, Vol. 3* (London: Oxford University Press) pp. 105–6, 636–7.

Somerville was not always easy to work with. Lance Sieveking, who worked alongside her in the Education Department in the mid-1920s, recalled that 'she was as much feared as she was loved' adding that it was said of her 'that she never went anywhere without creating consternation, havoc, a sensation or a precedent'.[69] She had a strong dislike, for instance, for Herbert Milliken, the Schools Executive, whom Somerville considered to be a fool.[70] She was also dismissive of a half-time assistant, Miss Simond who, because she was married and it was known wanted a child, was seen as insufficiently committed.[71] Somerville also developed a strained relationship with Professor Eileen Power who made regular school broadcasts in the mid-1930s.[72] Yet she was also much admired. Richard Lambert, recruited as an Education Assistant in 1927 before becoming Editor of *The Listener*, described the 'enterprising and self-possessed Mary Somerville' as 'one of the outstanding personalities in British Broadcasting'.[73] Roger Eckersley, who headed the BBC's Programmes Department from 1924 to 1934 and who was Somerville's boss for many years, was clear that school broadcasting was organised, administered and stimulated by her 'to whom all praise is due'.[74]

One person Somerville always got on with was Reith. From their first meeting in 1924 he saw in her the same zeal for public service broadcasting and commitment to education that he possessed. Reith's fondness for Somerville is conspicuous in his diary entries. Unlike senior male executives, about whom he frequently wrote damning entries, he never expressed any anger or frustration towards her. They often lunched and dined together; he personally congratulated her on her work and she was invited to visit him and his family at his country home, Harrias House.[75] Somerville's friendship with Reith also meant that she had privileged

[69] BBC/WAC:S61: Special Collections: Autobiographical Sketches of Lance Sieveking, p. 43.

[70] BBC/WAC:L1/305/2: Herbert Milliken Staff File, Nicolls conversation with Milliken, 12 June 1934.

[71] BBC/WAC:R13/216/1: Schools Broadcasting Department, document headed 'Schools Department Staff', 1 June 1933.

[72] Berg, *A Woman of History*, pp. 232–4.

[73] Lambert, *Ariel in All his Quality*, p. 54.

[74] Roger Eckersley (1946) *The BBC and All That* (London: Sampson Low, Marston) pp. 159–60.

[75] Reith Diaries, for example 15 February 1927, 22 January 1930, 12 December 1933, 29 May 1936.

access to him. Lambert recalled somewhat dryly that she could see the 'D.G.' at any reasonable time.[76]

Somerville's friendship with Reith was undoubtedly beneficial to her career. Her rapport with the all-male Control Board and her peers in senior management, all of whom were men (apart from Matheson), is harder to discern. Along with Matheson she was invited to the Control Board Tea, a routine instituted in 1930, whereby senior staff members were summoned to a broader executive gathering at least once a week.[77] However, unlike Matheson, Somerville left no record of what she thought about the men she worked alongside. Somerville was ambitious but not ruthlessly go-getting. She was not part of atmosphere of ambition, suspicion and intrigue that Maurice Gorham described at Savoy Hill.[78] While the men around her might have jostled for higher positions, Somerville was content to dedicate herself, firmly and calmly, to improving and expanding School Broadcasting. Lambert's description of her swimming round the Director of Education, Stobart, 'like a swan round a carp' is an apt one.[79]

Somerville was the only woman before the Second World War to sustain, long term, a senior management position in the BBC. She was retained despite her marriage and motherhood, evidence of her value to the Corporation and proof that it was possible for a woman to both care for a family and hold a high-powered job. Somerville is a prime example of a woman who carved out her own career at the BBC, a home-grown success for which she was rewarded in 1935 with an OBE. Her passion and commitment to school broadcasting set the agenda for the interwar years and drove the expansion of the department into the 1940s. Later she would inject this same drive to her job as Assistant Controller, Talks (from 1947) and ultimately as Controller, Talks (from 1950), the first woman to achieve this status. She retired in 1955. Reith's obituary of Somerville in 1963 is telling. Addressing her in male terms he wondered 'what this eager, restless, determined, irresistible pioneer, prince among men, and hero, is doing now'.[80]

[76] Lambert, *Ariel in All his Quality*, p. 54.
[77] BBC/WAC:R3/3/10: Control Board Minutes, 7 January 1930.
[78] Maurice Gorham (1948) *Sound and Fury* (London: Percival Marshall) p. 17.
[79] Lambert, *Ariel in All his Quality*, p. 54.
[80] *The Times*, 6 September 1963.

6.3   HILDA MATHESON (1888–1940)

'I am happier than anyone could believe possible', Hilda Matheson declared in June 1929, 'I have an ideal job and a very good screw and nice people to work with'.[81] Matheson's journey to the BBC had begun in May 1924 when, at the recommendation of Mary Somerville, she met John Reith.[82] He was evidently impressed by her abilities and the entrée that she might give the BBC to London's political and cultural circles, finally enticing her to the Company in September 1926 on a salary of £600.[83] According to Nancy Astor, Matheson was at first reluctant to take the post but Astor had insisted as 'hers was too good a brain to be kept only for my work'.[84] Matheson's initial BBC post, as an Assistant in Education, might sound unprepossessing but at this time the Education Department was responsible not only for school broadcasts, adult education and *Children's Hour* but also for religion, talks and news. Matheson joined at a propitious moment. Reith was in the process of reorganisation and a separate Talks Department was being formed, which came into operation in January 1927.[85] Whether Reith had earmarked Matheson as his Director of Talks is uncertain, Matheson clearly believed it had been Roger Eckersley's idea.[86] Eckersley, Director of Programmes and Matheson's direct boss, remained a great supporter of hers until the difficulties of the early 1930s led to friction and, ultimately, to her resignation in January 1932.

Matheson approached the job of Director of Talks with alacrity. The section she inherited was bland, timid and amateurish, she created a department

---

[81] HML, 11 June 1929. There is a large historiography on Matheson; see, for example, Carney, *Stoker*, Various Authors, *Hilda Matheson*; Briggs, *Golden Age of Wireless*, especially pp. 124–7, 141–3; Paddy Scannell and David Cardiff (1991) *A Social History of British Broadcasting, 1922–1939* (London: Basil Blackwood) particularly pp. 153–62. Matheson is also the focus of Fred Hunter, 'Hilda Matheson and the BBC, 1926–1940' in Sybil Oldfield, ed. (1994) *This Working Day World: Women's Lives and Cultures in Britain* (London: Taylor & Francis) pp. 169–74 and Charlotte Higgins (2015) *This New Noise: The Extraordinary Birth and Troubled Life of the BBC* (London: Guardian Books) pp. 17–36. Matheson also features heavily in Victoria Glendinning's biography of Vita. Victoria Glendinning (1983) *Vita: The Life of Vita Sackville-West* (London: Penguin).

[82] Reith Diaries, 27 May 1924.

[83] See, for example, Reith Diaries, 26 March 1926, 'Saw Miss H Matheson whom I think should be in the BBC somewhere'.

[84] Various authors, *Hilda Matheson*, pp. 15–16.

[85] While Stobart retained responsibility for Schools and Adult Education, Matheson became Director of a separate Talks Section.

[86] HML, 6 February 1929.

that was vibrant, challenging and professional.[87] For Matheson, broadcasting was about 'enlarging the frontiers of human interest ... widening personal experience and shrinking the earth's surface'. It was also about expanding democracy and fitting men and women 'for the complicated world of tomorrow'.[88] Matheson was committed to extending the scope of talks and bringing in the best speakers.[89] In 1927 a fledgling News Section was set up within the department, a response to the more flexible approach to news reporting instigated by the first BBC Charter.[90] She was in charge when the ban on controversial broadcasting was lifted in March 1928 which enabled the development of opinion pieces and more radical programming. For instance, in early 1928 listeners were introduced to Vernon Bartlett's *The Way of the World*, the first international series on the BBC. She also instigated broadcasts from the League of Nations Assembly in Geneva, the first time this had been attempted.[91] Matheson broadened the range of literary and critical talks and professionalised the way broadcast talks were given, insisting on careful scripting, lengthy rehearsal and tailored delivery.[92] She also greatly strengthened women's programming and brought increasing numbers of women broadcasters to the airwaves, as Chaps. 7 and 8 will show.

Matheson's approach to Talks reflected her liberal and progressive viewpoint. As part of London's cultural and intellectual elite her friends included not only the Astors but high-powered professional women such as the historian Marjorie Graves (also a Conservative member of Holborn Borough Council), the haematologist and radiobiologist Janet Vaughan (then at the start of her career) and Dame Rachel Crowdy (who headed the social questions section of the League of Nations). The most significant relationship of her BBC years was with Vita Sackville-West who she first met in December 1928 when Vita took part in a broadcast talk on 'The Position of Women Today'.[93] The encounter kicked off a two-year

---

[87] See, for example, Hugh Chignell (2011) *Public Issue Radio: Talks, News and Current Affairs in the Twentieth Century* (Basingstoke: Palgrave Macmillan) pp. 11–16.

[88] Hilda Matheson (1933) *Broadcasting* (London: Thornton Butterworth) pp. 14, 16.

[89] See, for example, Todd Avery (2006) *Radio Modernism: Literature, Ethics, and the BBC, 1922–1938* (Aldershot: Ashgate) pp. 45–50, 95–102.

[90] For an analysis of the significance of the News Section see Scannell and Cardiff, *A Social History of British Broadcasting* pp. 113–16; Hunter, *Hilda Matheson*, p. 171.

[91] Various authors, *Hilda Matheson*, pp. 22–6; HML, 3 September 1929.

[92] See, for example, Matheson, *Broadcasting*, pp. 71–7.

[93] The talk was with Hugh Walpole.

love affair, several months of which are documented in a series of evoca-
tive letters which paint a vivid portrait of life at the BBC.[94] The affair with
Vita enhanced Matheson's confidence at work. She knew Reith would
disapprove and she enjoyed the frisson it added to her daily life.[95] She was
known to be unconventional. Eckersley wrote that she was the only person
in the BBC allowed to bring a dog into the office and get away with it.[96]
She also conducted meetings on the floor around her fire which 'shocks
the great who may come in terribly', she told Vita.[97] How far these pec-
cadilloes were indulged because she was a woman is difficult to say, but she
enjoyed being different.

One of those whom Matheson wrote affectionately about in her letters to Vita was Lionel Fielden, whom she had personally appointed to
the post of Talks Assistant in 1927.[98] In his memoir, Fielden wrote how
initially he was uncertain that he could work under a woman, but he soon
changed his mind commenting that:

> Hilda was never preoccupied by power, never lectured, never laid down the
> law. She ran her department on a loose rein, encouraging, helping, sym-
> pathising and yet keeping herself firmly in the saddle.[99]

He also described her as 'one of those people who are made of pure gold
all the way through'. Matheson was held in high regard by most of those
she worked with. Richard Lambert claimed that, if it had not been for her,
he would never have accepted the job at the BBC.[100] Under Matheson's
leadership, he believed, the Talks Department entered a 'golden age'.[101]
The one person who was dismissive was Lance Sieveking who described her
as a 'busy little governess of a woman'. Sieveking, however, claimed that he

---

[94] The most intensive period of letter-writing was between 20 December 1928 and 1
March 1929, when Vita was in Berlin.

[95] HML, 3 January 1929.

[96] Eckersley, *The BBC and All That*, p. 100. The dog, a spaniel called Torquhil, had been
a gift from Vita.

[97] HML, 28 January 1929.

[98] She also wrote effusively about another personal appointee, Joseph Ackerley. 'Lionel
and Joe are dears', 'Lionel and Joe have been very sweet to me', HML, 14 January, 21
January 1929. Ackerley would go on to be Literary Editor of *The Listener*; Fielden to be
Controller of Broadcasting in India.

[99] Lionel Fielden (1960) *The Natural Bent* (London: Andre Deutsch) p. 114.

[100] Lambert, *Ariel in All his Quality*, p. 25.

[101] Lambert, p. 63.

was initially offered the post of Director of Talks only to have it rescinded in Matheson's favour, so this may well have coloured his view. [102]

Matheson was deeply committed to her job. In her book *Broadcasting* she wrote how broadcasters were never off duty and could never rest, being always on the lookout for new ideas.[103] This was certainly true for her; she rarely ceased working, whether it was reading manuscripts at home late into the night, rummaging through Vita's bookshelves for inspiration for poetry readings or courting politicians at lunch engagements.[104] Her working day varied enormously. It might be a series of long meetings: for instance, a Programme Board, a board to discuss controversy, a meeting with all the Station Directors.[105] It could be a succession of interviews with potential speakers: an Afghan, a docker, a man from the Royal Horticultural Society, four bridge players, a man from the music department to discuss combining poetry and music.[106]

Of all radio output in the interwar years, Talks caused the most apprehension because of the potential for controversy or offence. Roger Eckersley, as Director of Programmes, confessed that this was the most vulnerable and difficult side of his work. He dreaded 'the accidental passing of some statement with deep political or other implications which should have been blue pencilled and for which I should be responsible'.[107] As Talks Director, Matheson had to tread a fine line between what she described as the 'highly tendentious or the intolerably dull'.[108] There is no doubt that she often showed bravado, for instance, in January 1929 she arranged the first live debate to involve all three political parties, on the De-Rating Bill.[109] Reith and Carpendale were jittery. Carpendale, in particular 'got cold feet' but Matheson was reassuring and eventually had him 'wriggling on a pin'.[110] On the night of the debate itself, she rushed back to 'calm an agitated DG' but in the event it was a great success 'seventy minutes

---

[102] Sieveking, Autobiographical Sketches, p. 49. Sieveking left the Talks Department in 1928 to join the Research Department.

[103] Matheson, *Broadcasting*, pp. 51–2.

[104] For example, HML, 6 January 1929, 16 January 1929, 5 January 1929.

[105] HML, 21 December 1928.

[106] HML, 3 January 1929.

[107] Eckersley, *The BBC and All That*, pp. 156, 124.

[108] Matheson, *Broadcasting*, p. 93.

[109] The De-Rating Act was intended to encourage agriculture and industry, by freeing them from a portion of the rates.

[110] HML, 5 January 1929, 15 January 1929.

of it and it honestly wasn't dull'.[111] In some high-pressured situations, however, Matheson could lack self-confidence as her letters to Vita testify.[112] When she was worried, she doubted herself and became defensive, apprehensive that her gender might be seen as partly to blame; that she would be viewed as 'an unbalanced female unsuited to big jobs'.[113] The notion of inferiority, as we have seen, could be a major stumbling block for women. How different to the approach of Eckersley who, although 'uneasily conscious' of his limitations 'on the talks and educational side', always hid his weaknesses.[114]

In June 1929, Matheson reported her first major disagreement with Reith and Eckersley. She described to Vita an hour and a half's argument, 'hammer and tongs' on the future development of Talks.[115] Reith expressed anxiety that Matheson was straying too far into controversial ground; Matheson scorned Reith's opinions as those of someone ill-informed and little-read. Eckersley warned Matheson that she was getting a name for 'unreasonable truculence' which caused her to rail to Vita that they were 'always so damned ready to say to any *woman* who disagrees with them that it is unreasonable and shows a lack of balance'.[116] Although the quarrel was patched up she became convinced that there were plans to undermine her position.[117] In this she was justified. In October 1929 Adult Education Talks, under Charles Siepmann, were absorbed into the Talks Department, with Siepmann continuing to head his section and becoming second-in-command to Matheson. Siepmann, who had joined the BBC in 1927 as an Education Assistant, had initially been viewed by Matheson as 'a nice boy' but promoted to the position of a rival, their relationship faltered.[118] At the same time as Siepmann's promotion, Matheson lost control of 'Topicality' or Outside Broadcasts and News, which became separate departments. Topicality was an area she particularly relished and this was a further blow. In early 1931 there was another reorganisation, this time a separation of Talks and Adult Education, with Siepmann attaining equal

[111] HML, 22 January 1929.

[112] For example, she decried her 'damned thin-skinnedness', HML, 6 January 1929.

[113] HML, 20 February 1929.

[114] Eckersley, *The BBC and All That*, p. 137.

[115] HML, 20 June 1929. Reith commented in his diary, 'Saw Eckersley and Miss Matheson about her work and there is trouble brewing there', Reith Diaries, 20 June 1929.

[116] HML, 22 June 1929.

[117] HML, 28 June 1929.

[118] HML, 10 February 1929, 28 June 1929, 3 July 1929.

status with Matheson.[119] Matheson was seen to be undermined and her self-confidence again plummeted. This coincided with the end of her relationship with Vita who had begun a new love affair with Evelyn Irons.[120]

It was ultimately Matheson's relationship with Reith that was to be her undoing. Once weakened, she was unable to stand up to him and lost his respect. At first their relationship was congenial with many social occasions shared.[121] However, Reith's diary entry for 4 March 1930, recorded that he was 'developing a great dislike for Miss Matheson and all her works'.[122] The final straw came in the autumn of 1931when Reith and Eckersley attempted to water down Matheson's series *The New Spirit in Literature* presented by Vita's husband, Harold Nicolson. Nicolson was told he could not mention D.H. Lawrence and James Joyce and he threatened to pull out, pointing out the ludicrous nature of a programme on modern literature without reference to these two defining authors. Although a compromise was reached, Matheson felt unable to support the decision that there would be no allusion to *Ulysses*.[123] On 12 October 1931 she tendered her resignation to the BBC.[124]

News of Matheson's resignation reached the press in early December 1931 with much speculation as to the reasons for her departure. The *Manchester Guardian* hazarded the guess that it was because of differences of opinion between two opposing schools of thought 'one in favour of the intellectual type of talk and the other desiring a more popular note in the selection of talk topics and speakers'.[125] The *Evening News* described the difficulties of her job, trying to appease those who saw talks as trivial and dreary with others who were offended by anything controversial.[126] The headline in the *News Chronicle*, 'A Woman's Duel with the BBC'

---

[119] Matheson was Director of Talks (General); Siepmann, Director of Adult Education (Talks); Mary Somerville, Director of Schools (Talks).

[120] Evelyn Irons was the Woman's Page Editor of the *Daily Mail*. She had come to interview Vita on 6 March 1931. Glendinning, *Vita*, pp. 238–9.

[121] Reith Diaries, 'Miss Matheson has joined us and is doing well', 29 October 1926. On 1 March 1927, Reith recorded a lunch with Miss Matheson and Ernest Barker; on 13 October 1928, he went to Peaslake in Surrey with Miss Matheson to meet her family.

[122] Reith Diaries, 4 March 1930.

[123] Michael Carney drew extensively on Harold Nicolson's diaries for his understanding of Matheson's resignation from the BBC. Carney, *Stoker*, pp. 71–4.

[124] Reith Diaries, November 1931 (no date) Reith recorded receiving Matheson's resignation: 'It is her own fault that things have got to this pass, but she was quite mad about it'.

[125] *Manchester Guardian*, 4 December 1931.

[126] *Evening News*, 3 December 1931.

summed up the paper's opinion that it was her struggle with a management of men which had led to her resignation. However, there is little to support its view that Miss Matheson 'had pressed her views from a feminine standpoint in the face of overwhelming masculine opposition'.[127] Undoubtedly she was isolated in a man's world but the decisions she took were nearly always from an intellectual point of view. It was only when things went wrong that her gender became an issue, the fact that she was 'different' appears to have added to the tensions.[128]

While Matheson's resignation was partly personal it was also linked to the changing political climate, which viewed the BBC as increasingly left-wing.[129] A Conservative Government had been returned in 1931 and there was growing criticism of the Corporation in the right-wing press. As early as 1929, Matheson was writing to Vita about attacks on the BBC in the *Daily Mail*.[130] Whether Matheson actively promoted a 'left-wing' agenda has been widely debated as, by their very nature, talks by major cultural thinkers were often viewed this way.[131] Fielden admitted that he and Hilda attended a good many parties where intellectuals gathered. Hobnobbing with 'progressives', he claimed, was seen as tantamount to being a 'Red'.[132] However, there is no evidence that Matheson was a supporter of the Labour Party, indeed many of her closest friends were Conservative Party activists. Nevertheless, with hostility from both press and Parliament, Reith was anxious to stamp on what he perceived as left-wing bias in Talks. Matheson was identified with this and Reith was glad to see her go. Siepmann replaced her as Talks Director in January 1932.[133]

Was Matheson edged out because she was a woman?[134] It is true that men who fell foul of Reith, rather than leave, tended to be sidelined. Differences of opinion were tempered on the golf course or at the Club, a resolution unavailable to her. But Matheson resigned over principles.

---

[127] *News Chronicle*, 3 December 1931.

[128] Rosabeth Moss Kanter identified this as a factor that negatively affected women managers in the 1970s. Kanter, *Men and Women of the Corporation*, p. 214.

[129] Briggs, *The Golden Age of Wireless*, p. 141.

[130] HML, 20 December 1928, 27 December 1928.

[131] Scannell and Cardiff, *A Social History of British Broadcasting*, p. 155.

[132] Fielden, *The Natural Bent*, pp. 115–16.

[133] For a discussion on Siepmann's tenure as Talks Director see Scannell and Cardiff, *A Social History of British Broadcasting*, pp. 154–61.

[134] Michael Carney holds this view, Carney, *Stoker*, p. 79.

She was prohibited from carrying out her job in the way that she wanted to. Matheson never saw herself as part of the rat-race to the top of the BBC; she was not one of those pushing for power, which may have prompted her to leave rather than compromise. According to Fielden nine members of the Talks Department, including himself, were prepared to resign over her treatment, but she persuaded them otherwise.[135] We can only speculate as to where she might have ended up in the BBC hierarchy had she remained. Her successor, Siepmann, was similarly censured for being too radical (in 1935 he was transferred to the role of Director of Regional Relations), which suggests that Matheson's difficulties and the criticisms levelled against her were due to her outlook and the nature of the job rather than her gender.[136]

Following her resignation, Matheson maintained her links with broadcasting first as a radio critic for *The Observer* and *Weekend Review*, then as the author of *Broadcasting*, the first book to be written about the process of making radio, and ultimately as Director of the Joint Broadcasting Committee, a government-funded venture set up in 1939 which arranged for material about Britain to be broadcast by foreign radio stations.[137] Matheson had by then been diagnosed with Graves' disease. She did not survive an operation to remove part of her thyroid gland and died, aged 42, on 30 October 1940. At the time of her death she had lived with the poet Dorothy Wellesley for many years.[138]

6.4   ISA BENZIE (1902–1988)

Isa Benzie's arrival at the BBC in December 1927 was unpropitious. 'I thought her rather young and shy, and was surprised to find her 25', surmised Charles Carpendale following his initial meeting with her, but he added that 'Miss Banks liked her and thought she would do'.[139] The job was secretary to Major C.F. Atkinson, the BBC's Foreign Liaison Officer for which Benzie would earn £3 a week.

---

[135] Fielden, *The Natural Bent*, p. 117.

[136] See Scannell and Cardiff, *A Social History of British Broadcasting*, pp. 153–9.

[137] For a discussion on Matheson's career after the BBC, see Carney, *Stoker*, pp. 85–137.

[138] See Jane Wellesley (2008) *Wellington: A Journey through My Family* (London: Weidenfeld & Nicolson) pp. 289–90.

[139] BBC/WAC:L1/1049/2: Isa Benzie Staff File 2 (hereafter IBSF:2), Atkinson to Goldsmith, 29 November 1927, note from Carpendale.

During the War her father, Lt. Colonel Robert Marr Benzie, had served in the same army division as Reith, and it was through her father that Benzie first came to the attention of the BBC.[140] In June 1927 Col. Benzie wrote to Reith, at Isa's behest, informing his former comrade that she was 'a young lady of considerable character and ability'.[141] He praised her great linguistic ability, her musical prowess and the fact that she was currently training herself in shorthand and typing. He also emphasised her keenness to come to the BBC to do 'useful work', in particular something that would involve languages. Enclosing glowing testimonials from her College tutors, Col. Benzie concluded his letter with an exasperated stab at the young men who worked in his own office, wishing that they had 'half the industry and perseverance she possesses'. Reith's response to his old friend was positive. There were, he explained, periodically opportunities for girls like Isa, university graduates and people with considerable educational qualifications, but these were rare. Another possible approach was as a shorthand typist.[142] The process of her recruitment to the BBC was then set in motion.

The Foreign Department's role in the BBC was to maintain close and regular contact with broadcasting organisations in other countries. The global explosion in broadcasting in the 1920s and 1930s made it vital that the BBC was fully informed of all international developments and the Department acted as a clearinghouse for foreign activities. Benzie's temperament and skills were perfect for the job and she was soon taking over tasks from Atkinson, her boss, who had been redesignated Foreign Director in April 1928. Atkinson was highly impressed by Benzie. In October 1928, he pressed for a salary increase, on the grounds that she often stood in for him.[143] In March 1929, he remarked on her mental maturity, her high degree of initiative, her capacity for negotiating and her keen eye for policy implications in letters and minutes.[144] Because of this, he requested for her (and got) a wage rise to £5, the 'roof' of the clerical grades, a hefty increase of £1.10s a week. The following year, with Benzie now in effect 'the Executive of her Department', she was promoted to the monthly paid staff and Assistant grade on £300 a year.[145]

[140] Reith Diaries, 22 January 1924.
[141] IBSF:2, Colonel Benzie to Reith, 6 June 1927.
[142] IBSF:2, Reith to Colonel Benzie, 8 June 1927.
[143] IBSF:2, Wade to Carpendale, 28 October 1928. It was turned down on the grounds that secretaries were expected to deputise when their principals were on leave.
[144] IBSF:2, Atkinson to? (unclear), 6 March 1929.
[145] IBSF:2, Confidential Report, 1930.

Apart from managing the office, a key area of Benzie's work at this time was the organisation of relays from the Continent, the broadcasting of overseas programmes on the BBC. It was a complex procedure that involved detailed negotiations with foreign broadcasters to acquire the items, sorting out the timings of transmission, the publicity materials, copyright and so on. She was particularly praised by Atkinson for the success of a European tour in which she had arranged a series of open relays from Germany and Austria, 'a liaison which involved tact and command of an intricate situation' with foreign colleagues informing him that she created an 'excellent impression'.[146] In addition to her relay work, Benzie also entertained European callers, liaised with European broadcasters and frequently deputised for Atkinson when he was away.[147] Another of her roles was to provide information to Vernon Bartlett for his popular weekly BBC series *The Way of the World*.

In December 1932, the BBC's Empire Service was inaugurated. Although the Foreign Department maintained its distinctive role liaising with overseas broadcasters, it was deemed prudent to bring it within the new and larger Empire and Foreign Services Department under a new head, Cecil Graves. The following April Atkinson resigned (possibly he felt his position had been undermined), providing the opportunity for Benzie to assume his role. If the post of Foreign Director had been filled either by advertisement or by transference from another area of the Corporation, it would almost certainly have gone to a man. In the BBC of the 1930s, no woman was ever directly appointed to an executive post. As Foreign Director, Benzie's salary was to be £500 a year. Although this was a considerable rise for Benzie (whose annual earnings jumped by £150), it was markedly less than the £1,250 Atkinson had commanded.

To be Foreign Director was both a daunting and a prestigious position which involved negotiating at the highest level. The BBC Standing Instructions listed amongst the duties of the post: the collection, communication and record of incoming foreign information; executive dealings with foreign and international press, radio organisations and cultural movements; dealings with the public on foreign and international matters and entertaining foreign dignitaries.[148] The job also entailed straddling all foreign links initiated by other BBC departments, for example in Music and Education, and the Foreign Director was also called upon to represent

[146] IBSF:2, Confidential Report 1933.
[147] IBSF:2, Atkinson to Goldsmith, 6 February 1933.
[148] BBC Standing Instructions: Foreign Department.

the Corporation at meetings of the International Broadcasting Union (IBU) which frequently took place abroad. In May 1936, the *Morning Post* marvelled at Benzie's ability to 'ring up New York, Sydney, Calcutta or Cape Town as casually as you and I call a taxi'.[149] The same month *Wireless* magazine ran a feature on her, describing Benzie's work as 'a job that would intimidate many a man'.[150] In April 1937, *Ariel* reported on the Foreign Department's complex negotiations in connection with George VI's Coronation, a relay of which was being taken 'by practically every country in the world'.[151]

Like Somerville and Matheson, Benzie gave her all to the Corporation. In August 1935 she complained that she was working on average 14 hours a day. It was the only way she could complete even the basics of her job, she protested, and gave her little time for what she termed 'constructive work or initiation'.[152] She was often in correspondence with senior executives over low staffing levels and in September 1935 was finally successful in acquiring a second Assistant, Richard Marriot, which increased the full-time staff of the department to four.[153] The appointment of Marriot raises an interesting point. Why it was considered necessary for the new Assistant to be a man? In her plea for a second Assistant, Benzie had specified that 'he should be young, adaptable and presentable', with the 'ability to get on by himself'.[154] Graves was 'definitely anxious that the job should be given to a man', a point on which Benzie concurred.[155] The reason may be simple. With Janet Quigley already in place as an Assistant as well as a female shorthand typist, the only man working in the department in 1935 was the Foreign Executive, J.M.G. Best.[156] Perhaps it was felt that one more woman might have lowered the department's status. However, according to *Ariel*, in 1937, the Department now had a staff of eight, six of whom were female.[157]

---

[149] *Morning Post*, 18 May 1936.

[150] *Wireless Magazine*, May 1936.

[151] *Ariel*, April 1937.

[152] BBC/WAC:R13/206:Foreign Department, Benzie to Beadle (Empire and Foreign Executive), 27 August 1935.

[153] Richard Marriot joined Janet Quigley as an Assistant. Benzie had a personal secretary and also the full-time use of a short-hand typist in the General Office.

[154] BBC/WAC:R13/206, Benzie to Beadle, 27 August 1935.

[155] Beadle to Nicolls, 4 September 1935.

[156] Following the 1933 Production/Administration split all departments had an Executive, who acted as the liaison between Production and Administration.

[157] *Ariel*, April 1937. These included Benzie's 'personal clerk' Norah Wadsley and two further departmental secretaries.

Away from the office, Benzie functioned in a masculine world. At the BBC, the senior executives she worked with were all men.[158] She made frequent trips to Europe, as the BBC's representative at the IBU, again a male-dominated organisation.[159] At one point, Reith queried whether Miss Benzie should be given the same status accorded to Graves by the IBU, the implication being that, as Foreign Director and as a woman, she was of lower rank.[160] Perhaps, as a rare female, Benzie garnered more attention when she was abroad and in consequence was given greater consideration. Part of her job was to greet foreign dignitaries. At this level they would almost always have been men, yet the BBC appears to have been happy that one of its most important international ambassadors was a woman. It may have added cachet to an organisation keen to be seen as innovative to the wider world.

Benzie's work in the Foreign Department was applauded by her seniors, her rapid promotion to Foreign Director a result of her 'intelligence, wide knowledge and executive ability'.[161] In April 1936 it was announced that Benzie was to be given equal status with other Heads of Department in the Programme Division prompting pointed discussions about her comparatively low salary. While her manager, Graves, was satisfied to offer her annual increments of £50, which was in keeping with £500 a year, Nicolls, in his capacity as Director of Internal Administration, was adamant that her salary should be raised significantly and for three years in a row, a rise of £100 was authorised which was seen as more in line with the responsibilities of the job and in keeping with the role of a department head.[162] Benzie herself appears to have been unaware of these negotiations and there is no indication that she ever discussed her level of pay. In 1937, on the recommendation of her seniors, she received a further boost. She was to be regraded 'A' which meant not only enhanced status, but a new salary 'roof' of at least £1,250.[163]

---

[158] Cecil Graves as Empire Service Director; Lindsay Wellington as Director of Programme Planning and Basil Nicolls as Director of Internal Administration.

[159] See, for example, BBC/WAC:R3/3/11: Control Board Minutes, 14 January 1936, 17 June 1936, 7 September 1936.

[160] Control Board Minutes, 14 January 1936.

[161] IBSF: 1, Acting Administrative Assistant (Talks) to Administrative Officer (Talks), 7 December 1945.

[162] ISBF:1, Confidential Reports 1934, 1935, 1936. In 1936, the ten other department heads in the division earned at least £1000, with Val Gielgud earning £1,550 in Drama and John Coatman £1,600 in News. Mary Somerville earned £1,200.

[163] IBSF:1, Pym to Benzie, 2 June 1937.

However, Benzie's promotion in June 1937 coincided with the announcement of her engagement. Unlike Somerville, who was determined to work as a married woman, Benzie was 'clear in her mind' that she did not want to lead 'the double life that some girls do'.[164] On 2 September 1937 she married John Royston Morley, ten years her junior and one of the ambitious young producers in the new Television Department.[165] The event merited reports in several newspapers, where comment was made on it being a morning wedding (so allowing Benzie to return to work after lunch) and that only two witnesses attended, one of whom was Janet Quigley.[166] Benzie formally resigned on 3 January 1938, prompting genuine sadness at the loss of a woman of her abilities. *The Times* reported that she was given a farewell luncheon and presentation attended by Reith, Carpendale and the four Controllers, amongst many others.[167] Following Benzie's resignation, the appointment of a replacement was held up by the proposed reorganisation of Foreign and Empire Broadcasting. In the event, the position of Foreign Director was abolished. Marriot, one of Benzie's Assistants, resumed the title Foreign Liaison Officer in the newly amalgamated Home Intelligence Department.

It is interesting to speculate whether it was Benzie's resignation that prompted a restructuring of the department. In her four years in the post she had maintained if not raised its status, bringing it to a level with other department heads. Her personal style and diplomatic skills had given the job added prestige, but with her resignation the post ceased to be viable and became one of routine, reflected in her successor's title and lower salary.[168] Although Benzie had flair, she was not so large a personality as Matheson or Somerville (she was less vocal and demanding) and so made less of an impact on management. The post of Foreign Director was also not as high-powered as those of School Broadcasting Director or Director of Talks. Nevertheless, the importance of her role should not be underestimated; good foreign relations were vital to the interwar BBC and Benzie was often lauded in the press as one of those women who

[164] IBSF:2, Un-named memo to Nicolls/Pym, 7 June 1937.

[165] Reith was unimpressed by Morley. Meeting him for the first time he described him as 'a dilettante sort of youth, presumably brainy'. Reith Diaries, 5 May 1938. The marriage was not a success.

[166] *Daily Sketch*, 3 September 1937; *Daily Express*, 3 September 1937; *Radio Pictorial*, 17 September 1937.

[167] *The Times*, 5 January 1938.

[168] Marriot's salary was raised from £400 to £650.

held a significant job.[169] Benzie (who gave birth to a daughter in 1938), was to have a second BBC career. In 1943 she was re-employed as a radio producer, becoming the doyenne of health broadcasting as well as being instrumental in establishing the *Today* programme in 1957.[170] On her retirement in 1964, she was one of the few women in the Corporation graded 'A1'.

6.5   MARY ADAMS (1898–1984)

Mary Adams' initial encounter with the BBC was prickly. A respected scientist she had been invited to give her first broadcast talk in 1927. The proposed topic, 'The Differences between Men and Women' was, however, seen as potentially contentious, which necessitated a change of angle to 'Heredity and Environment'.[171] Luckily, the newly styled talk was 'extremely successful' and Adams was contracted to give a series of six further lectures on 'Heredity' the following spring.[172] Yet concern was again raised, this time about a reference to 'eugenists' (Adams' term) as well as allusions to birth control in her final script.[173] It was felt the statement 'knowledge of birth control methods must be spread where most it is needed' would bring controversy and criticism to the BBC, and so it had to be removed. After the series, while there were still some complaints from listeners that the broadcast programmes were too outspoken, most letters were highly appreciative. Adams' position as a provocative figure would be evident throughout her career at the BBC.

In January 1930, Adams applied for a permanent BBC post as Adult Education Officer for the Home Counties, her letter of application revealing that she had been intensely interested in this area of work since 1924.[174] Her list of affiliations was impressive and included membership of the British Institute of Adult Education as well as a seat on the executive

---

[169] See, for example, *Morning Post*, 6 September 1934; *Answers*, 5 October 1935; *Manchester Guardian*, 3 March 1936; *Express and Star*, 4 September 1936.

[170] Paul Donovan (1997) *All our Todays: Forty Years of Radio 4's 'Today' Programme* (London: Jonathan Cape) pp. 7–10, 18–19.

[171] BBC/WAC:S322/17/2: Mary Adams: Heredity, Lambert to Adams, 3 August 1927.

[172] Lambert to Adams, 6 October 1927.

[173] Lambert to Adams, 16 December 1927, 19 December 1927. For a discussion on the politics of birth control in the interwar years see Jane Lewis (1990) *The Politics of Motherhood: Child and Maternal Welfare in England 1900–1939* (London: Croom Helm) pp. 196–214.

[174] BBC/WAC:L2/5/3 Mary Adams Staff File:3 (hereafter MASF:3), 30 January 1930.

of the Eastern District WEA. Her letter ended with a statement of her belief in broadcasting as an educational tool and the 'immense influence on public opinion' that it could have. Adams was offered the job and joined the BBC on 5 May 1930, her £650 starting salary higher than any male colleague. She had originally asked for £800 but this was considered too high and instead she was persuaded to accept the lower figure, with the promise of £50 increments for four subsequent years.[175]

It quickly became apparent, however, that Adams was not suited to her new job. The work of an Adult Education Officer was largely organisational whereas she was an 'ideas' person. She became instead a programme assistant to Charles Siepmann who headed the BBC's Adult Education Section at this time. Her specialisation was science and she was soon producing well-received series such as *Pioneers of Health*, *Biology in the Service of Man* and *A1 or C3? The Future of the Race*.[176] Lambert considered her to be 'outstanding', certain that her 'contacts with scientists at the universities and her ability to pick out the latest scientific developments and have them presented in a lively and informative way' had transformed science broadcasting at the BBC.[177] Concerns remained, however, about Adams' impatience with office routine, her tendency to act without reference to seniors and her obvious frustration with a system that demanded her work be scrutinised by management; 'her enthusiasm is apt to outrun her discretion' cautioned Siepmann.[178] Although unspoken, her left-wing views were seen as potentially damaging to the Corporation at a time when they were desperately trying to prove themselves otherwise.[179]

As well as science talks for the Adult Education Department, Adams worked on General Talks, for instance the 1933 series *I Knew a Man* in which well-known figures looked back on the lives of great men and women they had known. She also produced debates on topical subjects; 'That Women Are Bored with Emancipation', broadcast in November 1935, being one example.[180] Her wide-ranging knowledge, her experience and

---

[175] BBC/WAC:L2/5/1: Mary Adams Staff File:1 (hereafter MASF:1), Goldsmith to Siepmann, 24 February 1931.

[176] For a discussion on Adams as a science producer see Allan Jones (2012) 'Mary Adams and the Producer's Role in Early BBC Science Broadcasts', *Public Understanding of Science*, 21, 968–83.

[177] Lambert, *Ariel and All his Quality*, p. 75.

[178] MASF:1 Annual Report, February 1931, January 1932.

[179] For a discussion on left-wing bias in Talks see Scannell and Cardiff, *A Social History of British Broadcasting*, pp. 155–60.

[180] The broadcast, on 2 November 1935, pitted E. Arnot Robinson against Nancy Astor.

her extensive network of contacts made her 'a very valuable member of the staff'.[181] Through her own connections and those of her husband, she knew an extraordinary range of people and her expense claims reveal a constant stream of dinners, lunches and sherry parties where she mingled with the great and the good.[182] But unfortunately for Adams, she was to take much of the blame for the infamous 'Ferrie Incident' of March 1934. William Ferrie, a representative of the National Union of Vehicle Builders, had been invited to participate in the series *The National Character*.[183] Representing the views of Labour, he was asked by Adams, not unreasonably, to redraft his highly political script.[184] Adams believed she had mollified Ferrie but on the night of the broadcast he announced that his talk had been censored and marched out, leaving a 30-minute gap to be hurriedly filled with gramophone music. The incident caused uproar in the press and for many months the BBC was reluctant to take political risks. Later, it was acknowledged that Adams had probably been blamed unfairly; the incident had occurred at a time of particular sensitivity for the Corporation and Adams was the obvious scapegoat.[185] However, it was considered so serious that there had been calls from senior management for her to be sacked.[186]

In the event, ill-health prevented this. Two years previously she had undergone a serious operation and was due a second operation in the spring of 1934. This meant that she was away from the BBC during the furore.[187] There was no way a paternalistic BBC could dismiss an employee at such a moment, especially as Adams had the support of her immediate managers and very probably, Reith.[188] Instead a compromise was found with Adams returning to a rare part-time job which removed the burden of too much office-based work. Although on reduced hours, she continued

[181] MASF:1 Annual Report, January 1934.

[182] For example, in May 1936 Adams claimed for 20 separate engagements. These included lunching Miss Buckley and Mr Rowntree (on the young architects group); consulting Sir Norman Angell (on the Peace series); sherry to Dr and Mrs Silk and Mr Roote (Prague journalists) and dining with Margery Fry (contact on Autumn programme), BBC/WAC:S322/39/2 Mary Adams Correspondence, Expenses May 1936.

[183] For more on the series and Ferrie's contribution see Scannell and Cardiff, *A Social History of British Broadcasting*, pp. 290–1.

[184] BBC/WAC:910FER: William Ferrie, letters from Adams to Ferrie.

[185] MASF:1, Nicolls to Graves, 30 April 1936.

[186] It looks probable that Colonel Dawnay, the Controller (Programmes), had instituted this.

[187] Although requested by Administration several times, no specifics of the procedure were given, the suggestion being that it was of a female nature.

[188] MASF:1, Nicolls to Carpendale, 19 July 1935.

to work on many complex programmes, for instance *Moonstruck Fish*, an exposé of the herring industry and another on the five-day working week. By 1935, Adams felt ready to return full-time, a decision supported by those she worked most closely with, who were now confident that she had tamed her political edge. Senior management, however, remained unconvinced.[189] For instance, at a meeting with Nicolls, the Controller (Administration) in August 1935, Adams insisted that the perception of her as 'a wild, unruly, Bolshevik sort of person' was unjust. Nicolls' response was that it was for her to prove that the view of her 'as being unruly and a political axe-grinder' was not a fair one.[190] By April 1936, the consensus was that Adams should return full-time, however, just as agreement was reached her pregnancy was announced.[191]

As had been the case with Somerville, Adams' maternity leave was dealt with in a pragmatic and business-like manner. It was to be generous. Reith agreed that she should get maternity leave as if she were on the full-time staff.[192] There was a suggestion that the option of working part-time should still be offered, 'Mrs Adams may prefer this herself when she has the baby, that is if she does not wish to retire altogether'.[193] Adams, however, was certain that she would want to return to work full-time.[194] On 28 September 1936, four weeks after her confinement, Adams wrote an effusive letter to Reith.[195] Thanking him for the lovely flowers and the welcome they gave to Sally, 'who is already in my uncritical eyes, quite adorable' she went on to enthuse that motherhood was very satisfactory, 'Already I feel a new creature, riding on ardour and responsibility—and considering how best I can justify my existence to my daughter'. Bearing in mind this new sense of purpose she informed Reith that she felt 'more and not less fitted for a position of responsibility' and, in consequence, she was keen to apply for the position of Director of Talks which had recently been advertised. Even today, to put oneself forward for such a major promotion a month after giving birth would be viewed as unusual. How much more so in 1936, but Reith hardly batted an eyelid and agreed to her application going forward.

---

[189] See, for example, MASF:1 Siepmann to Dawnay, 21 June 1935 , Rose-Troup to Dawnay, 25 July 1935.
[190] MASF:1, Nicolls to Dawnay, 19 August 1935.
[191] MASF:1, Rose-Troup to Graves, 7 February 1936.
[192] MASF:1, Reith to Graves and Nicolls, 21 April 1936.
[193] MASF:1, 20 April 1936 unclear who written by/to.
[194] MASF:1, Reith to Carpendale and Graves, 21 April 1936.
[195] MASF:1, Mary Adams to Reith, 28 September 1936.

However, he ended his letter with a caution, 'are you quite certain that you wish to live a double life, or are wise or right in doing so?'[196]

In September 1936, the BBC had finally publicised the position of Talks Director, a post that had been vacant for almost a year. Adams had expressed her interest in the job as early as April 1936, assuring Reith that she had 'very much changed her opinion of things since she joined the Corporation, and was much less revolutionary now than she was then'.[197] Adams, the only woman to be shortlisted for the £1,200 a year post, made a good impression on the Appointment Board.[198] The post, however, went to Sir Richard Maconachie who had served as British Minister to Kabul from 1930 to 1936. It was implausible that Adams would have been offered the job whatever her circumstances. With more than a thousand applicants, she faced stiff competition from outside and the BBC was already of the opinion that the position should be offered to someone of a right-wing persuasion. Maconachie's appointment satisfied this.[199] Adams was due to return full-time to the Talks Department in January 1937; however, visiting the BBC on his wife's behalf, Vyvyan Adams requested that because of stress, she be allowed a further few weeks' leave.[200] He also requested a move to Television for a few months, as he felt she needed a change from her present work. Whether Mary Adams was complicit in this request is not known, neither is it known why Television was the suggested destination. Probably the prospect of working under a new Talks Director was seen as untenable and Television Talks would have presented an exciting new challenge.

On 24 January 1937, Adams arrived at Alexandra Palace as the world's first woman television producer. The BBC had launched its pioneering television service in September 1936, which was available to a small number of London homes.[201] Adams was an unusual recruit both in her age (she was nearly 39) and in her sex, nearly all her colleagues were thrusting young men who had been specifically appointed to the new medium.

---

[196] MASF:1, Reith to Mary Adams, 5 October 1936.

[197] MASF:1, Reith to Carpendale and Graves, 21 April 1936.

[198] MASF:1, Carpendale to Adams, 3 December 1936.

[199] Briggs, *The Golden Age of Wireless*, pp. 148–9.

[200] MASF:1, Pym to Nicolls, 1 January 1937. Her father had died that week and she had just weaned the baby.

[201] For the early history of BBC television see Briggs, *The Golden Age of Wireless*, pp. 519–622. See also Gordon Ross (1961) *Television Jubilee: The Story of 25 Years of BBC Television* (London: W.H. Allen).

Gerald Cock, the Head of Television, welcomed Adams' wealth of contacts and ideas, but he was not impressed by her ability 'on the floor'. He judged her £800 salary 'seriously out of proportion' with others in the department and indeed Royston Morley, Eric Crozier and George More O'Ferrall, young men who would go on to have important television careers, earned considerably less.[202] Despite this, Adams continued to receive annul rises of £50 taking her salary in April 1939, to £900.

Adams brought to television her maturity, her experience and her brimming contacts book. Her responsibility was for talks, her output immediately impressive. She made her screen debut on 8 March 1937 introducing a programme on 'Architecture Today' with Maxwell Fry and Professor Walter Gropius and later that week a discussion on 'Food and Health' with Robert Hudson, MP and John Hilton, two men she knew well. She was also responsible for the output directed specifically at women, in particular the series, *World of Women* which included personalities such as the artist Dame Laura Knight, the sculptor Lady Kennett and the poet Olga Katzin, who spoke about their work. Adams produced general interest programmes such as *Picture Stories* where well-known artists told their life story through drawings, as well as a tribute to the charismatic priest Dick Shepherd, broadcast a year after his death.[203]

Described as small, birdlike and always well-dressed, photographs of Adams at this time capture her magnetism.[204] Frequently smoking a cigarette, she perches with her guests: purposeful, engaged, motivating. She was undoubtedly an intense woman and her early BBC career was marked by tension. But her value to the Corporation as a programme maker meant that concerns such as her political views, her ill-health and indeed her motherhood were surmounted by a management convinced of her ability and honourable in their behaviour towards a loyal member of staff. Like Somerville and Benzie, she would have an impressive post-war career with the Corporation promoted to Head of Television Talks in 1948 (where she worked alongside Somerville) and to Assistant to the Controller of Television in 1954, where she spotted the talent of Grace Wyndham Goldie.[205] Adams retired from the BBC in 1958.

---

[202] MASF:1, Confidential Report First Quarter 1938. In January 1937, Morley and Crozier, both Assistant Producers, earned £260; More O'Ferrall, a Producer like Adams, earned £400.
[203] *Radio Times*, 7 October 1938.
[204] *Oxford Dictionary of National Biography*, Mary Adams.
[205] Higgins, *This New Noise*, pp. 78–83.

6.6   CONCLUSION

The interwar years saw a small but increasing number of women attain elite positions as institutions, such as the Civil Service, were opened up and new business opportunities became within reach. The rise to executive positions, though, was still largely viewed as a novelty, as the tantalising newspaper stories of astonishing jobs and remarkable salaries attest. Mary Somerville, Hilda Matheson, Isa Benzie and Mary Adams were part of this much scrutinised band of pioneers. The four women had highly individualistic careers, nothing comparable to their positions existed outside the Corporation and within the BBC they also held unique roles.

There were some similarities between them, most obviously their graduate status. To have gained a place at Oxford or Cambridge was an immense achievement for a young woman at this time and imbued them with a confidence in their abilities that was evident at the BBC. Another similarity was their urbanity; they were modern-minded women who took the successes of feminism in their stride. A further trait they shared, and which was evident in other high-achieving women of the period, was their focus on the task in hand rather than seeing it as a stepping stone to something bigger. They were not part of the race-to-the-top mentality that characterised the careers of many professional men and which was especially prevalent at the young BBC. Because the women were not seen as a threat, it was easier for them to form positive relationships with those around them. All four women also held roles that were not perceived as feminine. Talks, foreign relations, science and television were areas of the BBC that, in future years, would be the domain of men. Schools management, although later feminised, was not viewed as such at this time.

Adams, Somerville and Matheson had rich hinterlands and were part of the cultural elite. They started their BBC careers in salaried roles, their intellectualism setting them apart from many of their male colleagues. Matheson, in particular, was unique in being the only woman who was head-hunted by Reith for an executive role. Benzie's trajectory was very different, the only one of the four who started at the BBC in a waged secretarial post. Obviously bright and ambitious, she was soon promoted, assuming the position of Foreign Director because she was perceived to be the natural successor. Somerville, too, grew into her role as Director, School Broadcasting. All four women were the right people, in the right place, at the right time. The pragmatic, and open-minded, BBC could see no reason to look elsewhere.

By 1939, Somerville was the only female Director still in post. Benzie was the last BBC woman to be appointed to a director-level position in the interwar years, after 1933, no further woman gained this status. In her biography of Eileen Power, Maxine Berg noted that the LSE, an institution which like the BBC took a progressive approach to the employment of women in the early years of the twentieth century, also experienced retrenchment.[206] Berg cited the growth of institutional constraints, more rigid hierarchies, increased professionalisation and the trend towards departmentalisation as possible reasons; all developments which were visible in the interwar BBC. Power was Professor of Economic History, a brand new subject area. She had been pivotal to its development and, initially, it had provided fresh opportunities for women.[207] Again, these were similar circumstances to the BBC. By the mid-1930s, the BBC had cultivated a sufficient number of home-grown male staff to supply most of its executive needs and when it was necessary or expedient to advertise jobs externally, it attracted a huge volume of interest from men.[208] Somerville was the sole woman to survive as a Director, and ultimately as a Controller, into the 1940s and 1950s. Adams would also have a significant post-war career but by then the nature of the Corporation and its attitudes towards women would have considerably changed.

---

[206] Berg, *A Woman in History*, pp. 180, 258–9.

[207] Berg, p. 8, 12.

[208] *Daily Express*, 1 October 1936. 'Yesterday was the last day for applications for the £1,200 post of Talks Director at the BBC. Just over 1,000 people applied for it, among them several women. I think it unlikely that the post will be given to a woman'.

CHAPTER 7

'When They Have Their Cup of Tea': Making Programmes for Women

In January 1938 Elise Sprott, the BBC's Women's Press Representative, gave a lecture to an international audience on 'Planning Broadcasts for Women in Great Britain'.[1] Here she spoke about the regular talks on cookery, child welfare, fashion and politics that were aired each week. However, she chose not to reveal that at one time she had herself been a producer of women's programmes. Four women in succession, Ella Fitzgerald (1923–1926), Elise Sprott (1926–1931), Margery Wace (1930–1936) and Janet Quigley (1936–1945), would have responsibility for the BBC's 'talks aimed at women' in the interwar years.[2] Hilda Matheson, during her tenure as Talks Director (1927–1932), would significantly develop and enhance the output, producing many of the talks herself. This chapter is about these five women and how their perception of what the BBC's female audience might want, coupled with their own interests and temperaments, would influence and shape the programmes they made.

Talks were a mainstay of the BBC schedules in the interwar years. Because production technologies were so rudimentary, a scripted and rehearsed 'live' broadcast was the cheapest and most straightforward way to deliver the spoken word. Florence Roberts had delivered the first talk

---

[1] The lecture, organised by the Associated Country Women of the World, was later included in a book. Elise Sprott, 'Planning Broadcasts for Women in Great Britain' in *Institute for Education by Radio* (1938) *Education on the Air* (Columbus, OH: Ohio State University Press). My thanks to Michele Hilmes for sending me a copy of the article.

[2] Women's talks were occasionally produced by male Talks Assistants, for instance George Luker.

© The Editor(s) (if applicable) and The Author(s) 2016
K. Murphy, *Behind the Wireless*,
DOI 10.1057/978-1-137-49173-2_7

aimed at women when she spoke on 'The Trends of Fashion' on 3 March 1923. Two months later, *Women's Hour* (not to be confused with the current *Woman's Hour*) was introduced. With the vast majority of married women based in the home, the fledgling BBC had quickly become aware of this captive daytime audience. Indeed, as Maggie Andrews has argued, the domestic setting of wireless and its association with the female sphere shaped a feminised form of address.[3] Although short-lived, *Women's Hour* would identify many of the conundrums that would beset the BBC's women's output into the 1920s and beyond. Should the talks be primarily domestic, or should they be a means of 'brightening their leisure hours' taking women outside the home?[4] Who was the woman listener? What time could she most easily listen? Should, actually, talks aimed only at women even exist?

The issue of tone was another challenge for the producers of women's talks, their audience might be predominantly housewives, but what was their level of education and sophistication? What sort of homes did they inhabit? What income might be assumed? Around 80 per cent of the UK's population in the interwar years were working-class and it was ultimately this audience that was principally addressed.[5] Yet, the lived experience of these women (which varied considerably) was largely alien to the programme makers themselves, all of whom were middle-class with a metropolitan outlook. The notion of 'uplift' dominated the interwar BBC—that middle-class culture and values were the stuff of radio and that this was what predominantly should be broadcast.[6]

The BBC's programmes for women exemplified its public service ethos. They were largely intended to inform and enrich women's lives, tapping into the enthusiasm for self-improvement and adult education that marked the interwar years. Whereas popular culture in terms of novels and the cinema might construct a dream world of wealthy Americans, rags-to-riches celebrities, Eastern sheiks and house parties, talks aimed at women on the BBC were either practical or edifying; there was little in the way of

---

[3] Maggie Andrews (2012) *Domesticating the Airwaves: Broadcasting, Domesticity and Femininity* (London: Continuum) pp. 3–7. See also, Kate Lacey (2013) *Listening Publics: The Politics of Listening in the Media Age* (Cambridge: Polity) pp. 113–26.

[4] *Radio Times*, 17 October 1924.

[5] Ross McKibbin (1998) *Classes and Cultures: England 1918–1951* (Oxford: Oxford University Press) p. 106.

[6] See for example David Hendy (2013) *Public Service Broadcasting* (Basingstoke: Palgrave Macmillan) pp. 7, 58–9.

fantasy or frivolous entertainment. In the USA, where commercial radio was driven by consumerism, the most popular daytime fare for female audiences was the soap opera.[7] The daily instalment of a gripping family-based drama ensured US women were hooked on sponsored programmes. From the mid-1930s, British audiences also had a new listening choice with the advent of continental-based commercial radio and during 1938 and 1939, for example, Radio Luxembourg and Radio Normandy broadcast English remakes of American soaps in the afternoons.[8] But despite this competition, the BBC continued to attract a large and growing female daytime listenership and, throughout the interwar years, there was determination that they should be well served.

7.1   Being a Talks Assistant at the BBC

In 1931, in a letter to Mary Agnes Hamilton, Matheson expressed her confidence in the quality of BBC talks:

> I do honestly believe that no other broadcasting service can show anything the least comparable in an attempt to cater for different audiences by hunting for appropriate speakers and ways of approach and insisting on the sorts of standard in both that we aspire to.[9]

The job of Talks Assistant at the BBC (the title given to talks producers from 1925) was multifaceted and involved, amongst other responsibilities, finding pertinent topics for broadcast, booking suitable guests, discussing with contributors the process of writing for radio, checking and advising on manuscripts, organising rehearsals and overseeing the final delivery of the talks in the studio.

Like all Talks Assistants, Fitzgerald, Sprott, Wace and Quigley were answerable to their managers in the Talks Department (or, in the case of Fitzgerald, what would become the Talks Department). Yet, because the programmes they produced were aimed at female listeners, they were not viewed as a high priority which meant they largely escaped the intense

---

[7] Michele Hilmes (1997) *Radio Voices: American Broadcasting, 1922–1952* (Minneapolis: University of Minnesota Press) pp. 154–75.

[8] Sean Street (2006) *Crossing the Ether: British Public Radio and Commercial Competition 1922–1945* (Eastleigh: John Libby) p. 249.

[9] BBC/WAC: Rcont 1: Mary Agnes Hamilton Talks File, Matheson to Hamilton, 29 August 1931.

managerial scrutiny of the prestigious evening talks. This gave the women a high level of autonomy and enabled them to adapt and develop their own interests and ideas for women's programmes, always within the Talks Department framework. The character and ethos of the Department changed during the interwar years, depending on who was in charge and the disposition of senior executives. Talks initially came under the umbrella of the Programme Director, Arthur Burrows and the Assistant Talks Director, Cecil Lewis before being moved to the Education Department headed by J.C. Stobart in 1924. Hilda Matheson, the first Director of Talks, assumed the post in January 1927 to be followed, on her resignation in January 1932, by Charles Siepmann. When Siepmann was transferred to the new post of Director of Regional Relations in 1935, the position of Talks Director was temporarily filled by J.M. Rose-Troup before Sir Richard Maconachie assumed the role in 1936. The left-wing–right-wing bias of the Department has long been a topic of academic debate (and was fiercely discussed in the contemporary press) with the Matheson/Siepmann years viewed as much bolder and more radical than those of Maconachie who was more conservative and more risk-averse.[10]

It was often pointed out that the BBC's female listeners were in the fortunate position of being able to enjoy their own dedicated programmes as well as the general talks output. An editorial in *The Listener* typified the view that, whereas women might often be interested in subjects intended for men, 'programmes designed specially for women will, with very few exceptions, interest women only'.[11] For this reason, the BBC rarely placed specific talks for women in the evening schedules. There might be women-related items, but these would be expected to have broader relevance. This did not mean that the daytime schedules were bursting with women's programmes. Music always made up the largest part of the broadcast day augmented by Schools output during term-times. But throughout the interwar years there was usually at least one 15-minute talk aimed at the female audience each weekday. Sometimes there were more, occasionally none at all, but they were a significant fixture.

Yet, as we shall see, there was concern that by targeting women the BBC might both pigeonhole its female audience and alienate its daytime male listeners, something that became particularly pertinent from the early

[10] See Paddy Scannell and David Cardiff (1991) *A Social History of British Broadcasting, 1922–1939* (London: Basil Blackwood) pp. 155–61.

[11] *The Listener*, 24 November 1938.

1930s when the Corporation embraced the issue of the unemployed.[12] From 1924, the notion of 'talks of general interest but with particular appeal to women' was the approach taken, which continued to be the over-riding policy of the interwar years.[13] After the demise of *Women's Hour* in 1924, these talks always had non-gender-specific titles such as *Morning Talks*, *Five O'Clock*, *Teatime Talks*. Even *Household Talks* avoided the use of the word housewife, although they were undoubtedly the intended audience.

The BBC, however, faced a far larger problem than how to frame its female-orientated output. Unlike the women's page of a newspaper or the abundant market in women's magazines, it did not have the luxury of a self-selecting readership; rather, apart from a national or regional option, its programmes were listened to by all.[14] As Mary Grieve, the Editor of *Woman*, pinpointed: 'pre-war divisions of taste and income set a strict limit on the number of like-minded women it was possible to gather together', hence the plethora of carefully targeted women's magazines.[15] As early as 1924, Fitzgerald was aware that on radio 'no such catholicity of choice is open to women listeners, since there is a common programme for all'.[16] But who was the woman listener? 'Cultured woman', 'young industrial woman', 'modern mother', 'Devon village woman', 'lonely spinster', 'Miss Mayfair', 'professional woman', were just a smattering of those imagined in *Radio Times*. The challenge of creating output that satisfied women of all ages and social classes; married and unmarried; homemaker and employee; in town and countryside was always going to be an almost impossible task. Understandably, most daytime talks reflected an audience that was predominantly one of housewives. Margery Wace, for example, would gain the sobriquet 'The Housewives' Friend'.

---

[12] See Scannell and Cardiff, *A Social History of British Broadcasting*, pp. 57–71.

[13] BBC/WAC:R6/219: Women's Advisory Committee, 30 April 1924.

[14] For a discussion on women's magazines and women's pages in newspapers see D.L. LeMahieu (1988) *A Culture for Democracy: Mass Communications and the Cultivated Mind in Britain between the Wars* (Oxford: Oxford University Press) pp. 39–43; Adrian Bingham (2004) *Gender, Modernity, and the Popular Press in Inter-War Britain* (Oxford: Clarendon Press) pp. 39–42; Fiona Hackney (2011) '"They Opened Up a Whole New World": Feminism, Modernity and the Feminine Imagination in Women's Magazines, 1919–1939' (Unpublished doctoral dissertation: University of London).

[15] Mary Grieve (1964) *Millions Made my Story* (London: Gollancz) p. 90.

[16] *Radio Times*, 17 October 1924.

It was not until 1936, with the advent of Listener Research, that the BBC was finally able to clarify that working-class housewives formed the bulk of the weekday listenership.[17] Margaret Bondfield MP would have concurred that it was this audience whose lives had been most radically changed by radio. Writing in *Radio Times* in 1937, she praised how 'the slender wire brings the world and its affairs into the tiny kitchens and living rooms which hitherto had isolated so many housekeepers in the performance of their duties'.[18] Matheson, too, was unequivocal that working-class women were the main benefactors of radio, citing evidence from 'particularly poorer women', as to what a wireless set meant to them as 'one of the most remarkable and encouraging results of broadcasting'.[19]

Matheson was also aware of a strong demand amongst women listeners 'for talks outside the common round of household drudgery, on travel, on books, on play producing … on how people live in other countries'.[20] Fitzgerald, Sprott, Wace and Quigley grappled with this balance between domesticity and edification throughout the 1920s and 30s, shaped by their understanding of modernity coupled with the social, political and cultural mores of the times. There is little to suggest that they were openly feminist in their outlook, rather they could be equated with what Alison Light has termed 'conservative modernity', bringing new ideas and ideals into the domestic sphere.[21] The influence of social maternalism is clearly apparent in the output, with a strong focus on childrearing, maternal health, infant welfare and diet.[22] The application of novel technologies such as gas and electricity along with expert and scientific ways of approaching housework are especially evident in the 1920s giving way to a more shared and experiential focus in the 1930s. Citizenship was always important, informing listeners about their rights and responsibilities, and was especially pertinent at the time of the 1928 Equal Franchise Act which extended the vote to all adult women.[23] New opportunities for women in

[17] Mark Pegg (1983) *Broadcasting and Society 1918–1939* (London: Croom Helm) p. 120.
[18] *Radio Times*, 12 November 1937, 'Women's Broadcasting Number'.
[19] Hilda Matheson (1933) *Broadcasting* (London: Thornton Butterworth) p. 188.
[20] Matheson, *Broadcasting*, p. 190.
[21] Alison Light (1991) *Forever England, Femininity, Literature and Conservatism between the Wars* (London: Routledge) pp. 10–11.
[22] Jane Lewis (1990) *The Politics of Motherhood: Child and Maternal Welfare in England 1900–1939* (London: Croom Helm) pp. 89–102.
[23] Michael Bailey has written about radio's place in reconciling women's role as housewives and mothers while at the same time recognising their enhanced function as citizens. Michael

terms of careers, sports, leisure pastimes and adventure were also strongly featured as were items on appearance and fashion.[24] As Adrian Bingham has shown, these were the bread-and butter issues of the popular press.[25] A sense of aspiration, of women's desire to improve their lives, pervaded the period and the BBC unquestionably tapped into this.[26]

7.2   ELLA FITZGERALD, 'WOMEN'S HOUR'

Mrs Ella Fitzgerald's arrival at Savoy Hill on 7 April 1923 coincided with the creation of a new programme, *Women's Hour*. The idea of Cecil Lewis, the BBC's Assistant Director of Talks, it was to be one of a trilogy of programmes, *Children's Hour*, *Women's Hour* and *Men's Hour*, that would be broadcast from 1 May to tie in with the opening of the eagerly awaited new studios.[27] While *Children's Hour* was already established, the distinct programmes for women and men were to carry items of particular relevance to these groups. *Men's Talks* (as the London programme was ultimately called) was short lived. Broadcast late at night, by October 1923 it had disappeared from the schedules, its diet of motoring, sport and whimsy deemed unsuccessful. *Women's Hour* as a distinct entity was also brief, but importantly it developed a style of women's programming that would continue throughout the interwar years.

Fitzgerald was one of the first women to be recruited directly to the BBC's salaried staff, indicating her status as a programme maker. Little is known about her life prior to the BBC apart from her place of birth, Dutch Guinea, and that she had worked in Fleet Street, at one time as a film critic for the *Daily Sketch*.[28] She was 35 when she arrived at Savoy Hill and it is likely that her maturity, coupled with her excellent contacts, made her a good choice to oversee *Women's Hour*, with its daily requirements for two talks, six days a week. Initially, Fitzgerald was to be Central Organiser for both *Women's Hour* and *Children's Hour* and each provincial

Bailey (2009) 'The Angel in the Ether: Early Radio and the Construction of the Household' in *Narrating Media History*, ed. Michael Bailey (London: Routledge) pp. 52–65.

[24] For a discussion on women and leisure see Claire Langhamer (2000) *Women's Leisure in England 1920–1960* (Manchester: Manchester University Press).

[25] Bingham, *Gender, Modernity, and the Popular Press*.

[26] Judy Giles (2004) *The Parlour and the Suburb: Domestic Identities, Class, Femininity and Modernity* (Oxford: Berg) pp. 47–55.

[27] BBC/WAC:CO9: BBCo Station Directors Meeting, 18 April 1923.

[28] *Ariel*, April 1936, October 1937.

station (of which there were now five), was instructed to appoint a Woman Assistant to supervise this output in their locality.[29] In November 1923, Fitzgerald was relieved of the responsibility for *Children's Hour* by the appointment of Geraldine Elliot to that role. This meant that she was able to give her whole attention to *Women's Hour*.

The first *Women's Hour* was broadcast at 5.00 pm on 2 May 1923 when HRH Princess Alice, Duchess of Athlone delivered the inaugural talk on 'The Adoption of Babies'. Also appearing on the programme was the famed couturier Lady Duff Gordon who spoke on 'Fashions'. This mix of the worthy and light-hearted, of domesticity, social issues and escapism would provide the blueprint for the programme. Amongst the regular topics introduced by Fitzgerald were cookery, poultry keeping, nursery chats, beauty, careers and bridge. She herself gave two talks a week: 'Ariel's Society Gossip' (presumably a round-up of society news) and 'In and Out of the Shops', a weekly update on what was worth buying, for which she used her lunch hour to gather material.[30] In order to stretch her programme budget Fitzgerald 'shamelessly exploited' former Fleet Street colleagues, several of whom came to the microphone 'once or twice, without fees, just for the novel experience'. Amongst those she tempted to Savoy Hill were the literary critic and feature writer Edith Shackleton, her sister Norah Heald of the *Daily Mail*, and Miss Hogg, Woman's Page Editor of the *Evening Standard*. Marion Cran, who gave her first gardening talk in August 1923, would become one of the first radio stars, as Chapter 8 will reveal.

The audience for the initial programmes would have been tiny, largely made up of women whose husbands were wireless enthusiasts.[31] The novelty of broadcasting, coupled with the small numbers of women listeners, would have meant that expediency was the main motivator for the choice of talks, Fitzgerald selecting subjects that she felt comfortable with and for which she could easily find speakers. Much echoed the standard women's fare in newspapers, with which Fitzgerald was familiar, with a strong focus on the domestic but also the changing reality of women's

[29] In April 1923 these were Manchester, Birmingham, Newcastle, Cardiff and Glasgow. Bournemouth and Aberdeen were added later, in October that year.

[30] *Prospero*, June 1969.

[31] Shaun Moores, (1988) '"The Box on the Dresser": Memories of Early Radio and Everyday Life', *Media, Culture and Society* 10(1), 23–39. By the close of 1923, 600,000 licences had been issued. Pegg, *Broadcasting and Society*, p. 7. A letter to Margaret Bondfield stated that she would be speaking to 'many thousands' of listeners, BBC/WAC: Margaret Bondfield Talks:1, Fitzgerald to Bondfield, 19 July 1924.

lives with talks on, for example, new technologies such as electricity and aluminium, new pursuits such as tennis and motoring and newly obtainable careers.[32] Reflecting on her time producing *Women's Hour*, Fitzgerald was palpably proud that the MPs Nancy Astor, Margaret Wintringham and Ellen Wilkinson had graced the airwaves and talks on citizenship were also prominent.[33]

In December 1923 a change was introduced to the way *Women's Hour* was run. A Women's Advisory Committee (WAC) was established to offer guidance to the programme, one of a number of advisory committees established by the early BBC to enhance its credibility.[34] Seven eminent individuals were invited to sit on the WAC, each signifying an area of women's lives the BBC deemed important to reflect. These were Lady Denman (Chairman of the National Federation of Women's Institutes), Margaret Bondfield (MP), Mrs H.B. Irving (the actress Dorothea Baird), Dr Elizabeth Sloan Chesser (physician), Mrs Violet Cambridge (Honorary Secretary of the Women's Amateur Athletic Association), Mrs Hardman Earle (Ministry of Food and Public Kitchens, First World War) and Evelyn Gates (Editor-in Chief, *The Women's Year Book*).[35] Within weeks of its first meeting two significant decisions were made; first to hold a poll of listeners' views and second to abolish the *Women's Hour* name.[36]

The plebiscite was Fitzgerald's suggestion. Two members of the WAC appeared on *Women's Hour* on 2 February 1924 to establish at what time women could most easily listen and which subjects were preferred. Mrs Hardman Earle put forward the case for 'practical talks on topics relating to the welfare of the home'; Miss Evelyn Gates supported the argument 'that women looked to the wireless as a potential means of brightening their leisure hours'.[37] Writing in *Radio Times* Fitzgerald described how the plebiscite had galvanised the listener, 'Keep us out of the kitchen!'

---

[32] In his study of women's interests in the popular press Adrian Bingham included housewifery, motherhood, consumer items, citizenship, fashion and 'modernity' which embraced new technologies as well as expanded employment opportunities, sport and women 'pioneers'. Bingham, *Gender, Modernity, and the Popular Press.*

[33] *Prospero*, June 1969.

[34] BBC/WAC:R16/219:Women's Advisory Committee. For more on Advisory Committees see Asa Briggs (1961) *The Birth of Broadcasting: The History of Broadcasting in the United Kingdom* Vol. 1 (London: Oxford University Press) pp. 240–50.

[35] Margaret Bondfield was unable to attend meetings and resigned in February 1924. Mrs Irving replaced Lilian Braithwaite who was also unable to attend.

[36] WAC Minutes, 18 January 1924, 25 January 1924.

[37] *Radio Times*, 17 October 1924, article written by Fitzgerald.

and 'Take us out of ourselves' was the overwhelming response with
75 per cent of responses imploring the programme 'to abandon at once
and for ever' all talks on domestic subjects.[38] 'Is it to be wondered at then'
Fitzgerald explained, 'that for "the cure of constipation" we substituted
a tour of Constantinople, that talks on the English country-side replaced
those on the stocking of the kitchen cupboard?' As a result of the plebiscite domestic subjects were reduced from one-third to a quarter of the
output, with a new time also agreed for the programme, 4.00 pm.[39]

Ironically, by the time Fitzgerald's *Radio Times* article appeared in
October 1924, *Women's Hour* had ceased to exist. At the February meeting, the WAC had unanimously agreed to abolish the name. No reason
was given as to why, but a distaste of special treatment for women was
part of the ongoing feminist debate of the mid-1920s. Divergent views
were expressed, for instance, by women's groups such as the National
Union of Societies for Equal Citizenship (NUSEC) and the Six Point
Group, centring on whether women should seek first to identify themselves as citizens rather than as women.[40] The outcome was that, from 24
March 1924, *Women's Hour* was no longer listed in the schedules. Instead,
from 4.00 pm, 'two talks of general interest but with particular appeal to
women' were interlarded with the afternoon concert.[41]

Despite the name being dropped, the WAC met twice more in April
and December 1924, and continued to influence women's programming.
Amongst recommendations, for instance, were a series of talks on *Psychology*
and on *The Domestic Service Problem* (both suggested by Sloan Chesser);
higher profile career talks by experts in the field; a broader range of travel
talks (which included a series on the Lake District by Elise Sprott) and talks
on hobbies by which listeners could earn pin-money. Later programmes also
included debates: 'That Woman is Nearer Barbarism than Man'; 'That the
Advantages of Education Are Grossly Overrated'; 'That the Eastern Woman
Is More Successful in Married Life than her Western Sister' amongst the
topics covered. There were also interviews with 'celebrities' such as the film
star Gladys Young, the novelists Rebecca West and Ruby Ayres and the

---

[38] Fitzgerald's interpretation of the figures was inaccurate. Out of the 326 letters received,
187 had voted against domestic topics, 57 per cent. WAC Minutes, 20 February 1924.

[39] WAC Minutes, 30 April 1924.

[40] See, for example, Cheryl Law (1997) *Suffrage and Power: The Women's Movement 1918–
1928* (London: I.B. Tauris) pp. 171–7; Olive Banks (1993) *The Politics of British Feminism,
1918–1970* (Aldershot: Edward Elgar) pp. 76–7, 135–6.

[41] WAC Minutes, 30 April 1924.

suffragist leader, Millicent Garrett Fawcett and with 'London characters'; a flower girl, a charwoman and a policewoman invited to talk about their lives. Much was innovative, pre-empting trends that would not be incorporated into general BBC programming for many years ahead. There was no indication that the December meeting would be the last but in September 1925 a letter was sent to all WAC members informing them that, as only one women's talk was now being given in the afternoon (rather than two), and as the title *Women's Hour* no longer existed, the need for an advisory committee to meet regularly had disappeared.[42] It appears that by getting rid of the name, the commitment to women's issues was gradually reduced.[43] Although Ella Fitzgerald's responsibilities may have diminished, she continued to offer an array of female-related topics such as 'Choosing a School', 'Psychology and the Shop Assistant', 'A Woman in the Wild—Tiger Shooting' and 'Holidays with a Car', eschewing, as before, an overabundance of housecraft.[44] Two of her regular broadcasters became so popular their talks were published as books: *My Part of the Country* by 'A Bonnet Laird' and Mrs C. Romanne-James's *O Toyo Writes Home.*[45] Fitzgerald's daytime listenership would have progressively increased during this time. By December 1926, the number of wireless licences holders in the UK had exceeded the two million mark.[46] Advances in technology, with improved reception and the use of loudspeakers rather than headphones, also meant that the experience of listening-in was becoming more palatable.[47]

In November 1926, Fitzgerald's three-and-a-half year association with talks for women came to an abrupt end when she was transferred to the new position of Assistant on *World Radio,* the foreign-programme supplement to *Radio Times.*[48] The move came within weeks of Hilda Matheson's arrival at the BBC who, in January 1927, began her tenure as Director of

---

[42] WAC, Letter to members of the Committee, 30 September 1925.

[43] In 1990, *Woman's Hour* campaigned to save its name for just this reason. Sally Feldman 'Twin Peaks: The Staying Power of BBC Radio 4's Woman's Hour' in Caroline Mitchell ed. (2000) *Women in Radio: Airing Differences* (London: Routledge) pp. 64–7.

[44] *Radio Times,* 18 May 1925, 20 September 1925, 19 February 1926, 8 June 1926.

[45] A. Bonnet Laird (1925) *My Part of the Country* (London: Herbert Jenkins) and C. Romanne-James (1926) *O Toyo Writes Home* (London: Herbert Jenkins).

[46] Pegg, *Broadcasting and Society,* p. 7.

[47] For a discussion on the development of wireless technology see Pegg, *Broadcasting and Society,* pp. 36–40.

[48] Fitzgerald would go on to become Assistant Editor in 1928, retiring as Overseas Press Officer in 1947.

Talks. Laying out her initial plans for the new department, Matheson made specific reference to the afternoon talks for women which she planned to bring more in line with evening talks, giving them an elevated status.[49] One of Matheson's first tasks was to replace Fitzgerald, and Elise Sprott, an established member of the Department, assumed the role.

7.3   ELISE SPROTT AND HILDA MATHESON, 'HOUSEHOLD TALKS'/'MORNING TALKS'

It seems that from the start Matheson was apprehensive about working with Sprott, whom she had not originally considered as Fitzgerald's replacement. Although there is little within surviving BBC documents to suggest that the two women had an uncongenial relationship (indeed it always appears to have been professional), Matheson's letters to Vita Sackville-West intimate otherwise. While Lionel Fielden and Joe Ackerley, Matheson's 'young men', are frequently lauded, Sprott is only mentioned once, described as 'fat Miss Sprott'.[50] Ultimately, tensions between the two would lead to Sprott being moved out of the Talks Department in 1931. Yet, despite the discomfort of their partnership, the four years they worked together would be critical to the development of women's talks, largely because their temperaments, and in turn what they brought to the programmes, were so different. Matheson was sophisticated, urbane and undoubtedly progressive in her views. Sprott, three years older, while enthusiastic and hard-working, was decidedly not an intellectual. She also appears to have been politically naïve. Where Sprott's interests lay predominantly in homecraft and the domestic sphere, Matheson was committed to widening the output, in particular enlightening women as citizens and, like many of her contemporaries, was an enthusiastic advocate of adult education.[51] These two approaches complemented each other: although the BBC had an important role to play informing women about the wider world, it was a fact that the majority of daytime listeners were domestically orientated. Sprott and Matheson represented these two different needs.

Within days of Matheson becoming Director of Talks, the afternoons were revitalised with a new regular slot for women, *Household Talks* which

---

[49] BBC/WAC:R13/419/1: Organisation of Talks Department, October 1926.
[50] Hilda Matheson Letters (hereafter HML), 11 June 1929.
[51] For example, Matheson was Secretary of the Joint Committee of Inquiry into Broadcasting and Adult Education 1926–28, chaired by Sir Henry Hadow.

was overseen by Sprott and was very probably her idea.[52] The scope of *Household Talks* was modest but practical. Broadcast on Mondays at 5.00 pm, the first few weeks included decorating a small flat, making a lampshade and luncheon and pancake recipes. Sprott ensured the talks were given by appropriate experts including Mrs Cottington Taylor, the Director of the Good Housekeeping Institute and Mrs Clifton Reynolds, an expert in household appliances whose own home was 'equipped with every modern convenience and labour-saving device'.[53] By the close of 1927, *Radio Times* was confident to report that there were 'few more regular audiences than those attracted by the *Household Talks*, for they are specially designed to be of practical value to their particular constituency'.[54]

Alongside *Household Talks* a second experiment was initiated in January 1927, a weekly schedule of programmes made in conjunction with the National Federation of Women's Institutes (NFWI) which Matheson had agreed to support.[55] Lady Denman, as Chairman, gave the introductory talk at 3.45 pm on Wednesday 12 January which introduced Rhoda Power's series *Village Life in Olden Times*. The NFWI collaboration continued throughout the year with several other series: *Citizenship in Practice, Health and Common Sense, Village Life in Other Lands* and *How to Improve our Villages*. A symbiotic relationship existed between the BBC and the NFWI at this time, with both institutions eager to promote education and citizenship amongst women.[56] Many Women's Institutes, for example, had installed wireless sets in their local village halls.[57] The NFWI supported the BBC's Listening Groups, part of the Corporation's Adult Education policy whereby numbers of like-minded people gathered together to listen to and discuss specially designated output which was supported with book lists and reading material.[58] In turn, for several years, the BBC covered the NFWI's

[52] BBC/WAC:R13/419/1: Talks Section Duties, October 1927.
[53] *Radio Times*, 2 September 1927.
[54] *Radio Times*, 30 December 1927.
[55] BBC/WAC:LE(E)1a: Contributors: Mrs Ray Strachey: 1a, Matheson to Strachey, 6 April 1927.
[56] For a social history of the NFWI see Maggie Andrews (1997) *The Acceptable Face of Feminism: The Women's Institute as a Social Movement* (London: Lawrence and Wishart).
[57] The Crawford Committee, established by the government in 1926 to examine the status of broadcasting, took evidence on this from the NFWI's Chief Organiser Mrs Nugent Harris.
[58] See Lacey, *Listening Publics*, pp. 140–9.

national conference and the organisation would also be a source of listener feedback and opinion.[59]

Besides these two established strands, other more general women's talks continued in the afternoons, including a number of long-running series, which would become a hallmark of Matheson's regime. For instance, in the autumn of 1927, the six-part series *The Growing Generation* included talks by eminent individuals such as Margaret Macmillan, originator of the nursery school movement Dame Janet Campbell, Senior Medical Officer to the Ministry of Health and the physiologist, Professor V.H. Motram. Part of Matheson's vision for radio was its role in democratising society and she was committed to bringing the most authoritative and prestigious speakers into the home. For instance, in 1928 Matheson approached Margaret Bondfield MP to kick off a new series *A Woman's Day* which profiled women in important civic roles.[60] Dame Katherine Furse who spoke about Jury Service on 28 November 1928 was a good friend of Matheson's, one of many who would grace the airwaves. The original intention had been to include a housewife and a factory worker amongst the speakers (Bondfield put the machinery of the Women's Section of the National Union of General Workers at her disposal for this) but as was the case with most of the talks Matheson produced, it was the expert voice that ultimately was aired.

Household Talks went from strength to strength, proving so popular that, at the request of listeners, a selection of talks were published as a BBC book, *Home, Health and Garden* and, from 24 August 1928, a designated women's page (with the same name) was introduced into *Radio Times*. For both of these, Sprott was the point of liaison. Sprott also conceived the idea of involving the listeners themselves in *Household Talks*. Audience participation in terms of requests and competitions had been commonplace on the early provincial stations and was a mainstay of *Children's Hour* but by the late 1920s, it was rare.[61] Sprott's suggestion

---

[59] For example, in July 1928, Matheson approached Inez Ferguson, the General Secretary of the NFWI, with a request that her members might 'report back on the suitability and acceptability of the talks'. BBC/WAC:R14/88: National Federation of Women's Institutes, Matheson to Ferguson, 11 July 1928. The NFWI annual meeting was covered on the BBC from 1929 to 1933.

[60] BBC/WAC: Margaret Bondfield Talks, 26 July 1928.

[61] In 1927 Matheson pioneered a late-night Prose and Verse Competition hosted by Naomi Royden Smith. See also Scannell and Cardiff, *A Social History of British Broadcasting*, pp. 309–11.

was that recipes and household hints could be contributed by listeners (which would be professionally read out on air) for which they would be paid a small fee.[62] An appeal for 'Listeners' Contributions' in *Radio Times* solicited more than 1,300 entries and the first *Listener's Household Talk* was broadcast on 24 September 1928 and monthly thereafter.[63]

In October 1928, bolstered by the success of 'Listeners' Contributions' and by the sale of more than 15,000 copies of *Home, Health and Garden*, Sprott approached Matheson with a new venture; a daily rather than a weekly household talk.[64] 'One weekly talk limits us very much in scope', she insisted, enthusing that 'there are a thousand and one subjects which present themselves'. The logical time for the new programme, Sprott believed, would be 10.45 am, 'when housewives are about their work and it should be early enough not to interfere with the shopping'. To place talks in the morning was a risk for the BBC. Apart from the *Daily Service*, only weather reports and gramophone music had been experimentally trialled at this time. Nevertheless, it was agreed that *Morning Talks* should go ahead, *Radio Times* hailing the 'new development' for early in the New Year as 'a special quarter of an hour for housewives and parents'.[65] On Monday 7 January 1929 'Law and the Home' was the first talk to be broadcast in the morning, on the Daventry 'Experimental' service.[66] Matheson was quickly convinced of the efficacy of the new 10.45 am slot, informing Reith of the impressive response from listeners and confirming her belief that 'the busy married woman who works at home' could listen more easily at this time.[67] By May 1929, *Morning Talks* were being broadcast nationally and would continue to be a fixture in the schedules until the outbreak of the Second World War.

Although Sprott had suggested this significant change to women's programming, Matheson strongly influenced the final content. Sprott had proposed conventional topics, for example on the principles of cooking, household art and the history of furniture.[68] While many of her ideas were

[62] BBC/WAC:R51/239: Household Talks (hereafter HT), Scheme for Special Series of Listeners' Contributions to Household Talks, 7 July 1928.

[63] *Radio Times*, 24 August 1928, 21 September 1928.

[64] HT, Sprott to Matheson, 2 October 1928.

[65] *Radio Times*, 23 November 1928.

[66] For more information on the BBC's transmission areas and experimental wireless, see Pegg, *Broadcasting and Society*, pp. 18–31.

[67] HT, Matheson to Reith, 7 February 1929.

[68] HT, Sprott to Matheson, 3 October 1928.

realised, when the new service came to air, far bolder series were also scheduled, initiated by Matheson. Ray Strachey's weekly *Woman's Commentary*
on social affairs, which will be examined in Chapter 8, was the first, airing
on Wednesday 9 January—another example of Matheson drawing on her
wide-ranging contacts in the political world. The following day, Margaret
Wintringham MP introduced the 15-part series *Our Boys and Girls* which
included such luminaries as Dr Letitia Fairfield on 'The Child at School'
and Vera Brittain on 'Careers'. The overtly domestic 'Menus and Recipes'
were the themes of the Tuesday and Friday talks, while on Saturdays,
talks encompassed lighter topics such as home decorating and fashions.
Alison Settle, the Editor of *Vogue*, who kicked off Saturday mornings with
'Who Makes the Fashions?' promised to take listeners 'behind the scenes
of the world in which fashions are made'.[69] Although publically supportive of *Morning Talks*, in private Matheson was nervous, declaring to Vita
Sackville-West ahead of the first broadcast, 'I blush to think how awful
you would think the whole lot of them'.[70] However, by mid-January, she
was greatly cheered by 'such nice letters ... coming from regular cottage
women', relieved that 'they do like the best ones best'.[71]

Morning Talks were predominantly uplifting and educational, expert
women, and sometimes expert men, imparting their knowledge into the
home. The themes also reflected many national concerns. For instance,
the series *Planning the Household Budget* broadcast in spring 1929, tapped
into the growing concern about the health of the nation and the renewed
emphasis on the mother's role in providing a better diet for her husband
and children.[72] The scope of the imagined listenership to the four talks
is evident in the programme's remit.[73] The first, on 8 April, considered
the family who lived on £500 a year. Mrs C.S. Peel, who presented the
talk, imagined this comfortably off family of four living in a modern villa
in a London suburb. Her second talk, 'How to Live on £300 a Year'

---

[69] *Radio Times*, 4 January 1929.

[70] Hilda Matheson Letters, 5 January 1929.

[71] HML, 16 January 1929.

[72] Anne Karpf has written about how BBC talks, from the early 1920s, were seen as an
important tool for health – Anne Karpf (1988) *Doctoring the Media: The Reporting of Health
and Medicine* (London: Routledge) pp. 32–3. Michael Bailey also stressed the significance of
the BBC in addressing women on issues of diet and health – Bailey, *The Angel in the Ether*,
pp. 57–9. See also Maggie Andrews, *Domesticating the Airwaves*, pp. 37–45.

[73] The talks were published each week in *The Listener*, 10 April, 17 April, 24 April, 1 May
1929.

pictured the same family structure in a five-roomed suburban, modern labour-saving home, but without domestic help. Margaret McKillop then broadcast two talks concerning the family on a weekly wage. The great difference between a working-class and a middle-class budget, McKillop indicated, was that the former never had sufficient money to get in stores or a place to put them if they could be afforded. She highlighted bread, margarine and milk as essentials, with eggs a good substitute for meat along with cheese and pulses. She was also prepared to accept that tinned food was a necessary aid to the overworked housewife. It is difficult to know who had the guiding hand in these talks, whether it was Sprott or Matheson. Mrs Peel would have been well-known to Sprott, having been one of the original broadcasters of recipes and kitchen news on *Women's Hour*, so it is probable that the programmes fell within her domain.

Hilda Matheson, on the other hand, was certainly responsible for *The Week in Parliament* (soon called *The Week in Westminster*), which began on 6 November 1929. Billed in *Radio Times* as, 'a series of weekly talks on the week's proceedings in parliament to be given by women MPs', the programme reflected the recent enfranchisement of all adult women.[74] One of the women Matheson hoped would take part in the series was Nancy Astor. In a letter to her former boss, she explained that the 'new experiment' was planned for 10.45 am 'the time when we find most busy working women can listen best, when they have their cup of tea'.[75] Informing women about politics was a priority for Matheson. During the run-up to the 1929 General Election, she ensured that women politicians representing the three major political parties were given airtime in the evening schedules.[76] She also developed the series *Questions for Women Voters*, broadcast fortnightly at 7.00 pm from October 1928, in which two adversaries debated contentious topics such as 'Should Married Women Work?', 'Should Women be Paid as Much as Men?' and 'Does Protective Legislation Benefit Women Wage-earners?'.

Sprott and Matheson continued to produce a wide range of talks, both in the mornings and afternoons. The future novelist Barbara Cartland, who had made her broadcast debut in 1926 defending the 'Youth of Today', became a Sprott regular starting with her *Morning Talk*, 'Making

---

[74] *Radio Times*, 1 November 1929.
[75] BBC/WAC: Viscountess Nancy Astor Talks:1, Matheson to Astor, 5 November 1929.
[76] Margaret Wintringham and Megan Lloyd George (Liberal) May 13, Margaret Bondfield (Labour) May 15 and the Duchess of Atholl (Conservative) May 17.

the Most of Oneself'.[77] Winifred Spielman's two talks on 'The Problems of Household Fatigue' (planned for some time by Matheson) were groundbreaking because for the first time listeners were invited to take part in a social survey.[78] Spielman, who worked for the National Institute of Industrial Psychology, was investigating fatigue in the home and hoped responses about, for example, washing up, laundry, mending and sewing, could lead to more scientific ways to carry out tasks more efficiently.[79] The series *The Future of Domestic Service* broadcast from January 1930 included broadcasts by grandees Lady Emmott and Violet Markham as well as one of Matheson's dearest friends, Kathleen Wauchope, which suggests that it was Matheson who had initiated the talks.[80] It also included the viewpoint of Miss Lizzie Willesden, a general servant, introducing a rare working-class voice to the airwaves.

In October 1930, Sprott instigated a further series aimed at homemakers, *Housewives News*.[81] The five-minute weekly consumer bulletin came to air in September 1931 and was both produced and presented by Sprott. However, by this time she had been moved out of the Talks Department to her new job as Women's Press Representative and Matheson herself was to resign shortly afterwards. There is a lack of clarity about Sprott's departure but it was undoubtedly linked to the arrival of Oxford graduate, Margery Wace. Matheson's letters to Vita show her dream of recruiting to the Talks Department a 'frightfully intelligent young woman of robust and excellent judgement', someone patently different to Sprott. At one point Vita suggested an approach to Oxford University which may have been how Miss Wace was found.[82] When Wace arrived in September 1930 she was initially to be an Assistant to Sprott.[83] Nine months later, in June 1931, the Control Board minutes recorded the ousting of Miss Sprott by Miss Wace, 'a more efficient junior'.[84] In the absence of the

[77] 18 November 1929 to 23 December 1929.

[78] The outcome of the survey is unknown.

[79] The final talk was published in *The Listener*, 8 May 1929. Christine Frederick, an American, was the doyenne of scientific management in the home at this time. Her book *The New Housekeeping: Efficiency Studies in Home Management* was published in 1913.

[80] Wauchope (referred to as 'KW') featured many times in Matheson's letters. For example, Matheson wrote that she owed her 'more than anybody else before'. HML, 30 January 1929.

[81] BBC/WAC:R51/241:Housewives News, Sprott to Matheson, 10 October 1930.

[82] HML, 28 January 1929, 11 January 1929.

[83] BBC/WAC:R51/646:Women's Programmes, undated document c.1931.

[84] BBC/WAC:R3/3/7:Control Board Minutes, 30 June 1931.

staff files of Sprott, Matheson and Wace it is impossible to know exactly what happened but Reith was evidently angered, commenting in his diary that there had been 'much discussion about Miss Sprott's case, both in general and in particular' and that he 'was very angry with the way she has been handled'.[85] He later added, 'I interviewed Miss Sprott who has been going through a difficult time, mostly Miss Matheson's fault', his rocky relationship with Matheson by this time starkly apparent.

7.4 Margery Wace, 'At Home Today'

Twenty-five-year-old Margery Wace was Matheson's perfect appointee. Not only was she Oxbridge educated but, prior to her arrival at the BBC, she had been secretary to the Oxford branch of the League of Nations and to the classicist and internationalist, Professor Gilbert Murray. Unlike cheerful Miss Sprott who could be unsophisticated in her dealings with contributors, Miss Wace was earnest and politically astute. For the first 18 months she worked alongside Matheson, who retained her interest in women's talks. However, after Matheson left the BBC in early 1932, Charles Siepmann, the new Director of Talks, handed over responsibility for *Morning Talks* to Wace, reporting the following year that they were 'developing rapidly into a service for women-in-the-home of great social importance'.[86]

Now sanctioned to impart her own vision, she refocused on the domestic in a manner she believed would both empower women and benefit wider society. Writing in *Radio Times* in 1935, Wace stressed her conviction that the mother in the home was crucial to social cohesion:

The nation's health, both physical and mental, is in the housewife's hands. By her skill and knowledge she must often make a very small sum provide adequate food; by sharing her family's interests, and by keeping her mind alert, she must provide a happy atmosphere in the home. We want to help her.[87]

The series in which this new conviction and approach can first be seen is *A Doctor to a Mother* broadcast from October 1932. Writing to Siepmann earlier that year Wace had suggested new arrangements for the child welfare talks which were 'potentially among the most important ... capable,

---

[85] Reith Diaries, end of June 1931 (no specific dates are given at this point in the diary).
[86] BBC/WAC:R13/419/2: Talks Department, Siepmann to Dawnay, 9 September 1933.
[87] *Radio Times*, 1 November 1935.

in time, of a real and lasting effect on the nation's health'.[88] Wace insisted that, because of this, not only should the most authoritative material be obtained but also the best speakers. She urged a change in the way they were chosen maintaining that, rather than depending on doctors in administrative positions, GPs should be used as not only did they have personal experience with child patients, but also used less jargon. The BBC had a 'better chance of being of value and help in these troubled times', she insisted, if it were free 'to approach the people who are doing work amongst children, and choose from them the best broadcasters'. The first talk of the series, 'Before Birth', given by an obstetric physician, tackled the worrying issue of maternal mortality which had remained stationary even though infant mortality had dropped.[89] In its billing *Radio Times* explained how 'It is hoped to improve this serious state of affairs, revealed by a recent Ministry of Health Report, by broadcasting the latest and best specialist information.[90] *A Doctor to a Mother* was soon augmented by other Friday morning series on child and maternal health such as *Common Sense and the Child* and *The Mother's Health*.[91]

In planning these programmes, Wace showed meticulous care. For example, in January 1934, for a series on the health of the school child, she wrote to the Matron-in-Chief of the London County Council requesting a meeting to discuss the problems facing school nurses.[92] Wace wanted her child welfare talks to tackle difficult issues such as stammering, shyness and obstinacy as well as more established subjects such as baby care. Her aim was to suggest to mothers how to build up healthy, happy children.[93] The Friday morning talks by doctors continued up until the Second World War. Hilda Jennings and Winifred Gill, in their 1939 report on listening habits in Bristol, made specific reference to them, believing they had a positive effect on the working-class housewives who listened.[94]

[88] BBC/WAC:R51/75: Talks: Child Welfare, Wace to Siepmann, 26 January 1932.

[89] Maternal mortality remained a worrying issue throughout the interwar years. See Lewis, *The Politics of Motherhood*, pp. 35–50, 121–5.

[90] *Radio Times*, 7 October 1932.

[91] Michael Bailey suggests these talks accorded women a new role as medical auxiliaries. Bailey, *Angel in the Ether*, p. 58.

[92] BBC/WAC:R13/419/2: Child Welfare Talks, Wace to Matron-in-Chief, County Hall, 30 January 1934.

[93] HT, Wace to Quigley, 7 April 1932.

[94] Hilda Jennings and Winifred Gill (1939) *Broadcasting in Everyday Life: A Survey of the Social Effects of the Coming of Broadcasting* (London: BBC) p. 17.

Wace's greatest innovation was to bring the voice of the housewife to the BBC. To hear 'real' people on air was not brand new. Since 1933, the variety show *In Town Tonight* had been presenting 'characters' (as had *Women's Hour* ten years' earlier) but in the context of Talks it was exceedingly unusual. In April 1934, the Talks Department, after receiving mixed reviews for two series on unemployment, had taken what was considered to be the brave step of inviting people who were out of work to come before the microphone.[95] The use of individuals with actual experience was an approach adopted to great effect by Wace, beginning with her series *How I Keep House* in which housewives themselves spoke about their domestic routine.[96] The programme was significantly different to previous *Morning Talks* series, such as *Planning a Household Budget*, which relied on the expert voice. In September 1934, *Radio Pictorial* ran an article headlined 'The Housewife's Friend' in which they wrote excitedly about Wace's journey around the country to meet potential contributors, 'So far, her search has taken her to Norfolk to see a farm worker's wife, to Scotland to visit a fisherman's and to Reigate to visit a policeman's home'.[97] The series was also enthusiastically previewed in *Radio Times*.[98]

To find her contributors, Margery Wace had indeed been conscientious, approaching organisations as diverse as the Amalgamated Union of Building Trade Workers and the Ministry of Agriculture. Writing to the General Secretary of the National Union of Railwaymen, Wace indicated that, in order to fit the series, the railwayman's wife she was seeking should have young children, family earnings of not more than £2 a week and the reputation of being a good housewife.[99] The notes sent to potential speakers show the very specific and personal nature of the talks, with the women being asked to portray the intricacies of their family set-up, the lay out of their flat or house and the amenities available to them. They were also to describe exactly what they spent their housekeeping on as well as the daily routine of their lives: what time they got up; how they approached the tasks of washing, ironing and sewing; where they shopped; how the arrangements for their children were made. Wace personally met all the contributors beforehand in their own homes, to assess

---

[95] See Scannell and Cardiff, *A Social History of British Broadcasting*, pp. 57–71.

[96] See also Andrews, *Domesticating the Airwaves*, pp. 46–50.

[97] *Radio Pictorial*, 21 September 1934.

[98] *Radio Times*, 17 August 1934.

[99] BBC/WAC:R51/240: Housekeeping, Wace to Marchback, 30 October 1934.

their suitability. This was important because, by definition, a working-class housewife would not be used to public speaking. Wace also oversaw their scripts and before each broadcast (which was live) the individual woman was extensively rehearsed. In her analysis of *How I Keep House*, Maggie Andrews suggests that through these broadcasts the women were given, momentarily, a national voice, their experience of domesticity the one that was defining.[100] Undoubtedly to allow ordinary women to broadcast, rather than experts, was a bold departure for the BBC and the series appears to have been well received. A congratulatory letter from a Scottish miner's wife, for instance, recounted that, despite running a home for nearly 19 years, the talks had taught her many useful hints.[101]

Using 'real' people as contributors was to become a characteristic of Wace's programmes, valuing the ordinary person's point of view as well as that of the expert. *Things I Remember* broadcast from September 1934 and *Would You Change?* from January 1935 were predicated on the memories and stories of women—and men—such as a kitchen maid from the 1880s, a window cleaner and an African Settler's Wife. Reflecting on the success of these programmes, Wace surmised it was because they had important psychological value for the listener. It made them feel that broadcasters 'were not only restricted to Prime Ministers and people of newspaper fame, but they themselves are also contributing to the programmes'.[102] Her championing of broadcasters such as Mrs Edna Thorpe, as Chapter 8 will show, was part of this innovative approach.

It was reflected also in a further development to the morning schedules in Autumn 1934, the introduction of *At Home Today*. Wace was keen to try something 'fresher' than the 'good works' that were characteristic of *Morning Talks*.[103] *At Home Today* was the BBC's first magazine-style programme, incorporating four short, diverse talks from both professional and amateur speakers; a new attempt to circumvent the constant dilemma of how to appeal to the breadth of audience. Topical with 'an element of surprise', was how the *Radio Times* billed the new programme, claiming that it would exclude 'nothing as inappropriate'.[104] An enthusiastic

---

[100] Andrews, *Domesticating the Airwaves*, pp. 48–9.

[101] BBC/WAC:R51/240, Letter from Mrs M Henry, 5 October 1934.

[102] BBB/WAC/R51/9/1: Adolescent Talks/Young Ideas, Wace to Siepmann, 22 March 1935.

[103] BBC/WAC:LE(E)1b: Mrs Ray Strachey Lectures: 1b, Wace to Siepmann, 20 September 1933.

[104] *Radio Times*, 6 September 1934.

reviewer in *Radio Pictorial* described how, in 20 minutes, they had heard 'the head of the Women's Police at Scotland Yard, a Spanish journalist, an ironmonger and an announcer reading an extract from [the Chief Medical Officer] Sir George Newman's report'.[105] The informal style of the programme may have contributed to its success. *At Home Today* was broadcast every Thursday until 1942.

One audience Wace believed was neglected when it came to the domestic was professional working women, and men, who could not listen in the mornings. In September 1935 she suggested a series of early evening talks, aimed unashamedly at the middle class, which could include more elaborate cookery recipes as well as tackling subjects such as house decorations, planning a labour-saving kitchen and choosing a prep school.[106] *Mainly Indoors* was given the go-ahead and from January 1936 the 12 talks, broadcast at 6.50 pm, included a varied mix of topics such as Children's Pets, Legal Problems of the Householder, Salads and Sandwiches from America, Understanding the Cars Complaints and a discussion on the One Maid Problem. The series appears not to have been a success, as it was not recommissioned, but it does indicate Wace's commitment to enhancing the domestic agenda regardless of the social status, or gender, of the listener.

In the spring of 1936, Margery Wace, along with Elise Sprott, organised the BBC Women's Conference, convened for 'an exchange of views on the subject of the BBC's Morning Talks'.[107] This was a major undertaking for the BBC, the first time it had ever arranged to meet with its listeners face to face.[108] Hosted by Sir Stephen Tallents, the newly appointed Controller of Public Relations, the Conference on 24 April 1936 was attended by almost 400 women representing more than 60 different organisations that included the Six Point Group, the Central Committee on Women's Training and Employment, the Mother's Union, the Women's Co-operative Guild and the Over Thirty Association. Twenty-six county federations of the NFWI were also present as were a number of eminent individuals.[109] Mary Agnes Hamilton and Lady

---

[105] *Radio Pictorial*, 4 January 1935.

[106] HT, Wace to Rose Troup, 18 September 1935.

[107] BBC/WAC:R44/86/1: Women's Conference 1936.

[108] In 1937 a similar event was staged for farmers who, like women, were identified as a discrete audience for whom specialist programmes should be made.

[109] The Conference Report stated that 397 women attended the morning session, 375 attended in the afternoon. In addition were special guests, press representatives and staff.

Caroline Bridgeman, the BBC's two female Governors, opened the proceedings which were also attended by senior executives from the Talks Department, as well as by Reith.[110]

The Conference, which was positively reviewed in the press, posed five questions: the timing of *Morning Talks* and whether there was a more suitable hour; whether the cookery and child welfare talks were of value; whether listeners liked regular talks on current affairs, in particular *The Week in Westminster*; whether there was a woman's point of view on subjects such as books and music and, if so, what approach was suggested and how the attending organisations might encourage their members to make more use of the talks. The timing of *Morning Talks* took up the greater part of the discussions, emphasising the differences and complexities of women's daily routines and demonstrating the insolubility of the conundrum of finding a period in the day that would suit everyone. When Sir Stephen Tallents, as chairman, asked for a show of hands, the audience were almost equally divided between 10.45 am and 2.00 pm.

Cookery and child welfare were affirmed as topics of interest and *The Week in Westminster* was also widely enjoyed, nevertheless almost every delegate pinpointed an improvement to the talks be it regional rather than London prices for ingredients, closer ties with welfare centres or a desire for speakers on local, as well as national, government. The discussion on 'Books and Music' also drew no consensus on 'a woman's point of view', instead broad issues were raised such as the listening needs of business women and a desire for talks on fashion and beauty culture that would interest the younger generation. Introducing the final session, Caroline Haslett, Secretary of the Electrical Association for Women, implored the BBC to continue the dialogue with women's groups that the Conference had instigated. In her closing remarks, Margery Wace made clear that a number of the suggestions had already been adopted by the BBC, a point emphasised by an exasperated Elise Sprott in the 1938 lecture which opened this chapter. Sprott acknowledged being startled to discover 'how little these good ladies … knew about what was already being broadcast'.[111] BBC executives were undoubtedly delighted with the Conference, reporting to the Board of Governors that it had been 'definitely a success'. But while it highlighted the Corporation's commitment to women's programmes,

---

[110] Reith apparently sat with his head in his hands for much of the time. Andrews, *Domesticating the Airwaves*, p. 51.
[111] Sprott, 'Planning Broadcasts for Women'.

there is little to indicate that many of the propositions put forward were formally taken up. Rather, the Conference underlined the impossibility of pleasing the diaspora of women who had vastly differing routines, habits and tastes.

The Women's Conference was one of Wace's last roles in the Talks Department. In August 1936 she was moved to the Empire Service, as Empire Talks Organiser, and was promoted to Empire Talks Director in 1941. Margery Wace died in 1944. She had married a BBC colleague, Ormond Wilson, in 1940; their daughter was born in 1942. Following the birth of a second daughter (who was stillborn) in 1944, she became mortally ill and died shortly afterwards.[112] *The BBC Year Book* included an obituary to her in which it emphasised how 'she [ensured] that in every broadcast talk for which she was responsible whether the speaker was great in reputation of a humble unknown, the virtue of the microphone should be exploited to the utmost'.

7.5   JANET QUIGLEY, 'TEA TIME TALKS'

Margery Wace's replacement in the Talks Department in August 1936 was 34-year-old Janet Quigley who was transferred from a demanding job as Isa Benzie's assistant in the Foreign Department where she had worked for six years. It was to be a shrewd choice, with Oxford-educated Quigley excelling as a talks producer.[113] Quigley would not only work on the output aimed at women but also on a number of high profile evening programmes including *Men Talking* and the controversial series *Towards National Health*.[114] Once ensconced in the Talks Department, she was quickly praised for her hard work, her ability to handle speakers, her good ideas and her discretion. The task of following Miss Wace was seen to be a difficult one, which she coped with admirably.[115]

Soon after Quigley's arrival in September 1936 there was a change to the schedules and a new experiment in women's talks, *Five O'Clock*. While this may have been a response to the Women's Conference it could also have been linked to the growing popularity of continental stations, such

[112] *The Times*, 14 January 1944. Wace was awarded an OBE for services to broadcasting in 1942.
[113] BBC/WAC:L1/784/1, Janet Quigley Staff File (hereafter JQSF:1).
[114] Quigley worked on the series with Guy Burgess. See Karpf, *Doctoring the Media*, pp. 37–8.
[115] JQSF:1, Confidential Report, April 1937.

as Radio Luxembourg, which had the effect of loosening up the BBC schedules from this time.[116] Broadcast weekly on a Monday, *Five O'Clock* replaced that day's *Morning Talk*. It is improbable that Quigley would have had an input into the change of time, as she had only arrived in the department the previous month, but she would certainly have influenced the content. Aimed at the family 'sitting round the tea table, waiting to be entertained' it continued the familiar pattern of a 15-minute talk by an expert, usually a woman.[117] In the nine months it was on air, this ranged from a motor-car consultant and the owner of a silkworm farm to an inventor, an antiques dealer and a golfing star discussing her wardrobe.[118] From September 1937, *Five O Clock* was replaced by a new series, *Tea Time Talks* planned on similar lines but, according to *Radio Times*, with 'perhaps ... a rather lighter note'.[119] The series, which boasted the first talk to women 'on making the most of their looks', survived until the outbreak of war.[120]

Quigley's zeal to help women is revealed in a fervent exchange of memos about a proposed new series in November 1936. The first memo, headed 'The Beauty Racket' disclosed Quigley's 'mission' to save women, particularly those who were badly off, from the 'tyranny' of beauty advertising.[121] She was particularly incensed, she wrote, by what she viewed as the conning of women into spending large sums of money on products with doubtful benefits, believing themselves doomed if they were unable to purchase the new face powder or anti-wrinkle cream. Her series, in contrast, would show women how they could keep, 'skin, hair and hands in good condition and also indulge in moderate cosmetics for ten shillings a year!' Because this was a potentially contentious topic, Quigley needed approval from senior management but the response from Cecil Graves, Controller of Programmes, was a curt 'no'. Having discussed the idea at Programme Committee there was concern that the fury of manufactures would be raised if the BBC was seen to suggest that women could make themselves beautiful by soap and water alone. In addition, from the point of view of *Radio Times* advertisements, which included beauty products, it would be difficult.[122] Not to be deterred, Quigley re-presented her

---

[116] See Scannell and Cardiff, *A Social History of British Broadcasting*, pp. 230–5.
[117] *Radio Times*, 28 September 1936.
[118] Broadcast 19 October 1936, 9 November 1936, 4 January 1937.
[119] *Radio Times*, 23 September 1937.
[120] In fact Barbara Cartland had spoken on this topic in 1929.
[121] BBC/WAC:R51/397/1a: Talks Policy, Quigley to Rose-Troup, 9 November 1936.
[122] BBC/WAC:R51/397/1a, Graves to Rose-Troup, 9 November 1936.

'considerably modified' ideas to Maconachie, the new Director of Talks, in July 1937. Assuring him that she was no longer 'burning with indignation' at the beauty trade, she hoped that an occasional series of talks on the care of the skin, hands and hair might be included in the autumn schedules.[123] The series was again proposed to Graves, Maconachie citing as his excuse 'the quenching of Miss Quigley's spirit'.[124] This time it was agreed, and Mary Embrey presented the first 'Making the Most of your Looks' as part of *Teatime Talks* on 21 October 1937.

As well as overseeing the new afternoon programme, Quigley took over with enthusiasm the Friday morning child welfare talks. While appreciating their value, particularly for working-class mothers, Quigley was anxious about the way correspondence was handled.[125] Because the doctors used in the series were anonymous, letters from listeners were not forwarded on to them for reply which, she believed, severely reduced the value of the service. Writing to Maconachie, Quigley gave the example of a mother of three, expecting her fourth, who had written to the BBC following a programme on the importance of diet during pregnancy. 'Yes, she knew that she should drink two pints of milk a day at this time, but what was the use of telling people like her when she had not enough money to get herself an extra cupful, let alone a pint'.[126] By not responding, Quigley claimed, the BBC were withholding much needed information that free milk was available at the nearest health clinic. Quigley's suggestion, that at the end of each series an entire programme should be devoted to clarifying information and answering queries, was taken up.

Another of the responsibilities Quigley assumed was *The Week in Westminster*. Under the stewardship of Wace the programme had undergone a major modification when, in 1931, male MPs were introduced as speakers. In January 1937, with a new time-slot was proposed, Quigley sensed that the character of the programme was again about to change, voicing her regret as she believed it provided an important service for women.[127] Maconachie confirmed that there was indeed an intention to broaden the talks beyond the female audience. It needed to appeal to a larger section of more serious-minded listeners and from September 1937

---

[123] BBC/WAC:R51/397/1a, Quigley to Maconachie, 6 July 1937.
[124] BBC/WAC:R51/397/1a, Maconachie to Graves, 14 July 1937.
[125] BBC/WAC:R51/210/1: Health, North Regional Programme Director to Maconachie, 1 February 1937.
[126] BBC/WAC:R51/75, Quigley to Maconachie, 19 April 1937.
[127] BBC/WAC:R51/115: Week in Westminster, Quigley to Maconachie, 7 January 1937.

The Week in Westminster was divorced from women's talks.[128] Quigley, however, insisted on producing the new-look programme which she did until it was temporarily discontinued at the outbreak of war.

In April 1937, Quigley instigated the first of her own series, the experiential *Other Women's Lives*. Broadcast at 10.45 am on Saturdays, it brought to the microphone an eclectic assortment of expert and 'ordinary' women including Agnes Smith, who began her working life as a 'doffer' in a cotton factory, Mrs Edward Harvey who ran a general store in a working-class district of Liverpool, the film critic Winifred Holmes and Mrs Olga Collett, supervisor of women staff for ICI, whom we shall meet in Chapter 8. Like her predecessor, Quigley's output predominantly addressed women as homemakers with series such as *Sickness in the House* about the care of invalids and the elderly and *Housewives and Experts* aimed at those who did their own DIY.[129] A six-part series on *Careers for Girls* in spring 1939, made in collaboration with Ray Strachey and the Women's Employment Federation, was constructed as a short talk from a 'young and enthusiastic' member of each selected job (nursing, physical training, domestic service, dressmaking, the Civil Service and secretarial work) followed by a brief description of the training required, promotion prospects and so on which were provided by Strachey.[130] Quigley had requested that the talks be broadcast in the early evening 'so that fathers as well as mothers could hear them' but in the event, they were placed at 2.45 pm on Saturdays.

The series *Mistress and Maid* broadcast from January 1938 revisited a perennial interwar issue: the servant problem.[131] The shortage of domestic staff had been a preoccupation of the middle classes since the end of the First World War with the employment of a maid continuing to symbolise affluence. The topic had been addressed within women's talks several times before, both in individual programmes and in broader series.[132] Quigley

---

[128] BBC/WAC:R51/115, Green to Barnes, 24 June 1937.

[129] BBC/WAC:R51/397/2: Talks Policy:2. The reports are dated 29 April 1938 and were sent by Quigley to Maconachie.

[130] BBC/WAC:R51/69/4: Careers, Quigley to Maconachie, 1 February 1939.

[131] The shortage of servants was viewed as a significant problem for the middle classes, see for example Alison Light (2007) *Mrs Woolf and the Servants: The Hidden Heart of Domestic Service* (London: Fig Tree); Lucy Delap (2011) *Knowing their Place: Domestic Service in Twentieth Century Britain* (Oxford: Oxford University Press).

[132] In 1924, 'The Domestic Service Problem' had included broadcasts by women who employed servants and servants themselves as had the 1930s series 'The Future of Domestic Service'.

replicated the format of the earlier broadcasts which had utilised a range of viewpoints; in her series, 16 different speakers were used. As with previous series there was no debate about servitude itself rather, as an article in *The Listener* elucidated, those who spoke had expressed opinions on why the servant shortage had arisen and how best the situation might be dealt with, stressing issues of status, flexibility and pay.[133] Quigley was satisfied the subject had been given a thorough airing and was particularly pleased that in many households, mistress and maid had listened together, discussing the talks afterwards.[134] The correspondence had also been impassioned, hailing the BBC on the one side, 'as the courageous spokesmen of a maligned and inarticulate class', and, on the other, 'accused, usually by contented servants, of stirring up unnecessary trouble'.

Janet Quigley's final daytime programmes, before the BBC introduced its wartime schedules, increasingly focused on the looming hostilities.[135] 'The Housewife in an Emergency', for example, broadcast on three consecutive Tuesday mornings from 15 August, included advice on food shortages, tips on storage and the use of canned food. This gave a foretaste of Quigley's extensive wartime series for women which would include *The Kitchen Front, Calling All Women, Your Health in Wartime* and *Talking it Over*, for which, in 1944, she was awarded an MBE.[136] In 1950, after a break of five years following her marriage in 1945, Quigley would resume her association with women's programmes as Editor of *Woman's Hour*, retiring as the BBC's Assistant Head of Talks in 1962.[137]

7.6   CONCLUSION

The BBC broadcast a wealth of talks aimed at its women listeners during the interwar years. That these were produced by women is not a surprise. The journalists who worked on the women's pages of newspapers and as the editorial staff of women's magazines were nearly always female, perceived to be more closely attune to their readership and to the woman's point of view. But whereas women journalists were usually cogs within a

---

[133] *The Listener*, 23 February 1938.
[134] Talks Policy: 2, Janet Quigley Quarterly Reports, 29 April 1938.
[135] This was introduced on 4 September 1939.
[136] See Sian Nicholas (1996) *The Echo of War: Home Front Propaganda and the Wartime BBC* (Manchester: Manchester University Press) pp. 70–100.
[137] See Kristin Skoog (2010) 'The "Responsible" Woman: The BBC and Women's Radio 1945–1955' (Unpublished doctoral thesis: University of Westminster).

much larger team, those who produced women's broadcasts at the BBC generally worked alone and with a high degree of autonomy. This was because daytime talks aimed at female listeners were not a high priority for the BBC so, provided they did not stray into areas of controversy, the women who produced this output tended to be spared intense managerial scrutiny. Ella Fitzgerald, Elise Sprott, Margery Wace and Janet Quigley were dedicated to producing programmes they believed would enhance women's lives: whether a talk on the uses of electricity, a sponge pudding recipe, an exploration of infant health or the revelations of a shop girl's life.

Although seldom light-hearted, talks for women fitted the Reithian principles of the BBC: to inform, educate and entertain. They addressed their daytime audience essentially as homemakers and mothers, offering insights on health, diet and infant care, but also providing stimulation and escape. As Hilda Matheson noted, radio offered women 'a preparatory course to help them to catch up, to feel less at a disadvantage, to keep abreast of wider interests'.[138] In her role as Talks Director, Matheson brought a greater professionalism and consistency to women's programmes, introducing coherent series and a more regular timeframe. She replaced Fitzgerald's often exuberant and somewhat ad hoc approach with a more 'highbrow' vision; a determination to bring to the female daytime audience the best and most up-to-date in expert opinion. While Matheson's particular focus was citizenship, Sprott championed the domestic and this continued to be the main thrust of women's talks during the remainder of the interwar years. Wace and Quigley brought heightened rigour and professionalism and a closer engagement with the housewife, who herself began to appear on air.

The 16 years from 1923 to 1939 were years of rapid change for women in Britain. For the BBC, the sense of awe at the burgeoning possibilities for women at the beginning of the 1920s became, by the decade's end, an urgency to inform them of their new obligations as voters and citizens. The early 1930s saw a heightened awareness of women's role in making-ends-meet which had settled, by the mid-1930s, into a lighter, more reflective mood before gearing up at the end of the decade to the new realities of war. The BBC's women's programmes were not about radically altering women's lives. They rarely questioned women's place in the home which was accepted as the norm. Neither Fitzgerald, Sprott, Wace or Quigley

[138] Matheson, *Broadcasting*, pp. 189–90.

exhibited overtly feminist tendencies. They were realists with a commitment to what they believed would improve women's lives. The output entwined the safe and old-fashioned with the modern and style-setting and unquestionably introduced new ideas, new practices and new experiences into the home. From within the confines of their kitchen or living room, women were able to get the latest expert advice on childcare, to travel to distant lands, to learn about the lives of others, to hear for themselves the opinions of female MPs, scientists and academics. Chapter 8 now moves on to the women who entered the nation's homes as they came before the microphone, as broadcasters, commentators and announcers.

'You Feel Their Personal Touch':
Women Broadcasters

To listen to some of the women who have broadcast is to forget immediate limitations ... it seems, as you sit listening, that what is spoken into the
transmitting microphone is a message sent specially to you. The millions of
other hearers are forgotten. The voice just whispers to you. It is a little *tête-
à-tête* in the company of the great.[1]

This extract from 'What Women Listeners Gain', a 1926 *Radio Times* article by Lady Alexander, evokes an essential quality of radio, its intimacy.[2]
The 'private' way in which wireless was consumed in the home, in the parlour, in the living room, in the kitchen, raised many issues about how the
audience should be addressed.[3] At first the significant difference between
radio and public speaking and the written word was not fully appreciated by the BBC. 'Anyone who sounded as if they had done something
colourful in their lives' was plucked from *Who's Who* and given 'almost
unrestricted freedom of the ether'.[4] But gradually it was realised that to

---

[1] *Radio Times*, 1 January 1926.
[2] See, for example, Martin Shingler and Cindy Wieringa (1998) *On Air: Methods and
Meanings of Radio* (London: Arnold); David Hendy (2000) *Radio in the Global Age*
(Cambridge: Polity Press); Hugh Chignell (2009) *Key Concepts in Radio Studies* (London:
Page).
[3] See, for example, Maggie Andrews (2012) *Domesticating the Airwaves: Broadcasting,
Domesticity and Femininity* (London: Continuum) pp. 3–26.
[4] Ralph Wade (undated) *Early Life at the BBC* (Unpublished memoir) p. 3. Wade was the
first Programme Assistant employed by the BBC, joining a few days before Ella Fitzgerald.

© The Editor(s) (if applicable) and The Author(s) 2016   221
K. Murphy, *Behind the Wireless*,
DOI 10.1057/978-1-137-49173-2_8

be successful at the microphone required not only an engaging topic and a certain level of authority but also an understanding of the personal nature of the medium, the knack of an appropriate script and the facility to deliver it convincingly. Women would form an important part of this pool of adept speakers. However, issues about their voice and questions about when, and on what, it was acceptable for them to talk would create tensions both for the women themselves and for the programme makers who worked with them.

Mary Somerville, newly recruited to the BBC, is credited as the first person to have requested changes to a script and insisted on a rehearsal.[5] This was in December 1925 for which Somerville was berated for impertinence by her manager J.C. Stobart (the broadcaster in question was the 'distinguished Professor of Literature', Oliver Elton). Hilda Matheson as Talks Director streamlined the process creating the leaflet 'Broadcast Talks and Lectures: Suggestions to Speakers' in December 1927.[6] Here, both 'The Manuscript' and 'The Voice' were considered, with advice to speak slowly and to avoid monotony; to write as you would talk to a friend; to refrain from using long sentences and difficult words; to grab with the opening phrase an audience of 'every kind of mentality and every degree of education' and above all, not to declaim. 'The note of *intimacy*, as in conversation with a friend, is the note to adopt', the leaflet stressed. To be a broadcaster in the interwar years required not only the ability to speak appropriately but also, for the BBC, the certainty that the individual would not offend, that what they said was accurate and true (or was clearly their opinion) and that they were tasteful and respectful in their manner.

It was Matheson who changed the culture of broadcasting at the BBC, enticing the nation's political, cultural and social grandees to Savoy Hill. Many were women and her tenure as Talks Director was a comparatively rich one for female broadcasters, as this chapter will show. Thousands of women would come before the microphone in the interwar years and while this was predominantly in the daytime—on School Broadcasts, on *Children's Hour* and on the output aimed at women—they were also a small but influential presence in the evening schedules. Most of those who spoke on the radio did so just once or perhaps occasionally; it was only those who were invited to Savoy Hill and Broadcasting House on a

---

[5] Mary Somerville's BBC Swansong, Governors Dinner, 20 December 1955.
[6] BBC/WAC:R13/419/1: Talks Department, 'Broadcast Talks and Lectures: Suggestions to Speakers', 22 December 1927.

regular basis who gained the title 'broadcaster', a new career created by the BBC. Their recognition was reliant not simply on their ability to write a strong script and to engage the audience but was also dependent on their 'expertise' as well as on the relationship they developed with their producer, as a focus on four women—Marion Cran, Ray Strachey, Beatrice Webb and Edna Thorpe—will reveal. Cran's garden expertise, Strachey's wide-ranging cultural experiences, Webb's status as a pre-eminent social scientist and Thorpe's 'ordinariness' as a housewife gave each woman a particular authority to speak. The voice that personified the BBC in the interwar years was that of the announcer, and also that of the eye-witness commentator, roles held almost exclusively by men. Two women, however, would make their mark in these areas: Sheila Borrett who became infamous for her failure and Olga Collett who became famous for her success as this chapter will show.

8.1   BEING A WOMAN BROADCASTER ON THE BBC

Lionel Fielden, who joined the BBC as a Talks Assistant in 1927 (and who was viewed as a progressive) made the stark claim that women were 'almost never good broadcasters' declaring 'I don't know why this should be, but it is a fact'.[7] Fielden acknowledged that famous women were not neglected by the BBC but in the area of Talks he asserted, 'you can reel off lists of men who have been, or who are, stars of the microphone: but you will have a job to find any women who equal them'. And, with the most popular voices quickly established as regulars, it is the likes of Sir Walford Davies, S.P.B. Mais, Commander Stephen King-Hall, Desmond MacCarthy, A.J. Alan, John Hilton, C.H. Middleton and Harold Nicolson who are best remembered, and who made their mark in the evening schedules of the BBC.

The BBC always prized its evening talks most highly because not only was the audience large but this was when men predominantly listened. Alongside lighter and more general talks, this was where the big social and political issues of the day were vented, always under intense scrutiny from politicians, notables and the press.[8] Unsurprisingly, it was male

---

[7] Lionel Fielden (1960) *The Natural Bent* (London: Andre Deutsch) p. 111.

[8] For impartiality and controversy in Talks see Paddy Scannell and David Cardiff (1991) *A Social History of British Broadcasting, 1922–1939* (London: Basil Blackwood) pp. 23–38, 153–78; Asa Briggs (1961) *The Birth of Broadcasting: The History of Broadcasting in the*

broadcasters who dominated these talks. Most required authority and expertise and it was men who held the vast majority of Britain's senior positions. In addition, most BBC's Talks Assistants were male and when drawing on personal networks to find speakers, they tended to choose men. There were some notable exceptions. For example in March 1929 the economist Barbara Wootton presented the series 'Some Modern Utopias' while the industrialist and newspaper editor Viscountess Rhondda was invited to speak as part of the 'Wither Britain' series in 1934 alongside such luminaries as Churchill, Ernest Bevin, George Bernard Shaw and Lloyd George.[9]

As early as 1924, Reith had made clear his view that 'a man should be of pre-eminent and recognised position if he is to speak to the whole country'.[10] The BBC's mission to improve the nation—'uplift' as it was frequently referred to—entailed the presentation of middle-class cultural values as well as a middle-class voice. As LeMahieu has pointed out, the individuals of social standing or intellectual eminence who were invited to the microphone were perhaps more important for who they were than what they said.[11] It was largely Matheson's ability to persuade the cultural and political elites of the virtues of broadcasting that cemented her position of eminence at the BBC. Her daytime programming also made wide use of high-powered and expert women who imparted the cream of their knowledge and experience into the home. The notion that authority might lie in experience as well as expertise was gradually appreciated by the BBC. The use of the 'plain man' ('or plain woman!' as Matheson declared) was first considered in 1929.[12] By the mid-1930s, Margery Wace was taking great pains to find ordinary women to take part in her *Morning Talks* and Olive Shapley also pioneered social documentaries that, for example, included miners' wives.[13] Almost without exception, however, those who

United Kingdom Vol. 1 (London: Oxford University Press) pp. 253–75; Asa Briggs (1965) *The Golden Age of Wireless: The History of Broadcasting in the United Kingdom,* Vol. 2 (London: Oxford University Press) pp. 124–60.

[9] Lady Rhondda complained about the lack of women broadcasters in the 1930s. Angela V. John (2013) *Turning the Tide: The Life of Lady Rhondda* (Cardigan: Parthian) p. 525.

[10] John Reith (1924) *Broadcast over Britain* (London: Hodder and Stoughton) p. 148.

[11] D.L. LeMahieu (1988) *A Culture for Democracy: Mass Communications and the Cultivated Mind in Britain between the Wars* (Oxford: Oxford University Press) pp. 182–3.

[12] BBC/WAC:R51/118:1a: Talks: Debates and Discussions 1926–36, Matheson to Eckersley, 20 August 1929.

[13] Olive Shapley (1996) *Broadcasting: A Life* (London: Scarlet Press) p. 53.

gave regular talks on the BBC had Received Pronunciation. In 1926, the BBC established its Advisory Committee on the Spoken Word with an expert in phonetics as its chair. The BBC's potentially important role in the standardisation of speech meant that the norm agreed on was 'educated English' and the predominant voice would be that of the Southern English middle and upper class.[14]

The quality of a woman's voice could, however, be viewed as problematic, one of the supposed difficulties being pitch. Speaking at a conference in January 1938, the BBC's Women's Press Representative, Elise Sprott delivered the 'sad truth' that women's voices 'on the whole are not awfully good over the microphone'. The reason, she explained, was the vibration of the voice, 'the soprano voice does not broadcast very well, and unfortunately most women are soprano'.[15] Matheson was also aware of this practical problem, observing in 1933 that it was 'true that most women's voices do not yet transmit as well as men's with their lower register'.[16] A letter to *Radio Times* in 1939 berated the BBC for not giving women broadcasters a fair hearing, the present microphone built on the experience of men. 'Engineers, give the women a microphone that will respond to the lighter and more rapid vibrations of their voices' the writer implored, 'and we shall find what they have to say interesting and instructive'.[17] Whether microphone technology in the late 1930s was still causing issues is a debatable point, but it was seen to be an issue.

While there were evidently troubles with the 'mike', the character of a woman's voice could also cause concern. An article in *Radio Pictorial* in April 1937 took women to task for sounding 'swanky and condescending'.[18] Filson Young, the BBC's official programme advisor, was also critical of women broadcasters who 'seem to err on the side of over-assertiveness'.[19] An exasperated letter-writer to *Radio Times* in 1932 pinpointed two women broadcasters for particular opprobrium: 'one had

---

[14] See Scannell and Cardiff, *A Social History of British Broadcasting*, pp. 176–7; Mark Pegg (1983) *Broadcasting and Society 1918–1939* (London: Croom Helm) pp. 160–2. Hendy argues that Received Pronunciation was viewed as an equalising force, David Hendy (2013) *Public Service Broadcasting* (Basingstoke: Palgrave Macmillan) p. 24.

[15] *News Chronicle*, 29 January 1938, The Conference was organised by the Associated Country Women of the World.

[16] Hilda Matheson (1933) *Broadcasting* (London: Thornton Butterworth) p. 56.

[17] *Radio Times*, 23 June 1939.

[18] *Radio Pictorial*, 2 April 1937.

[19] *Radio Times*, 16 November 1934, 'Women's Broadcasting Number'.

what I should call a serpent complex, and the other is surely a reincarnated buccaneer'.[20] BBC Talks Assistants, both male and female, were often dissatisfied with the performance of the women they enticed to broadcast. The novelists Rose Macaulay and E. Arnot Robinson were 'extremely disappointing'; the journalist Vera Brittain 'weak and ineffective', although dissatisfaction was also extended to men.[21]

Yet dozens of women broadcasters did gain approval from the listening public. Vita Sackville-West, Maude Royden, E.M. Delafield, Mary Bagot Stack and Ann Driver (the creator of *Music and Movement)* are just a few who were praised in letters to *Radio Times.* Olga Collett, the BBC's first woman Outside Commentator, was also held up for acclaim, as will be explored. Mary Agnes Hamilton, another popular speaker who broadcast regularly on books, politics and current affairs, took umbrage with those who did not like women's voices. In her forward to the *Radio Times* 'Women's Broadcasting Number' in November 1934, she felt confident that this would soon change, citing the 'sound sense' of Lady Reading, Professor Winifred Cullis and Eileen Power.[22]

The experience of broadcasting changed as wireless and studio technology advanced and production methods matured. Looking back on her appearances at Savoy Hill in the mid-1920s, Elise Sprott conjured up an era when the microphone was hung on a moveable stand 'neatly covered with a blue or pink shade' and when 'comfortable chairs and desks were still things of the future'. She had stood to broadcast, allowing 'each leaf of the script to flutter gently to the ground'.[23] Lady Cynthia Asquith was less relaxed, 'My Ordeal at the Microphone: How it Feels to Broadcast' was how *Radio Times* headlined an article by her in November 1924.[24] Here she recalled the trepidation of walking down Savoy Hill, the lozenge swallowed too hastily in the lift, the arrival in the 'torture chamber' of the vast studio where she was asked to sit and wait her turn. Finally summoned to the microphone, she tried desperately to remember the guidance given as to the speed she should read her script, fearful that if this should be either too fast or too slow, the schedule would be thrown into disarray.

[20] *Radio Times,* 5 February 1932.
[21] BBC/WAC:R51/118: 1a, 13 March 1936, Adams to Rose-Troup; Vera Britain Talks, 23 August 1940.
[22] *Radio Times,* 16 November 1934.
[23] *Evening News,* 30 June 1936.
[24] *Radio Times,* 21 November 1924.

Fortunately she managed her ten-minute allocation exactly before she was politely thanked and directed out into the street.

Cynthia Asquith's fee is not recorded but the usual fee for an evening talk was ten guineas. Daytime speakers were paid less (five guineas for a standard talk, two to three guineas for a short talk in *At Home Today*) although individual rates were sometimes agreed. The MP Margaret Bondfield, for example, agreed a sum of eight guineas for an afternoon talk, the contracts executive certain that she would appreciate that the BBC 'were unable to offer fees for talks in the afternoon as high as those for talks in the evening'.[25] From 1929, *The Listener* included edited versions of talks for which broadcasters were paid an extra fee, although only occasionally were these selected from the daytime output. Matheson, who that year arranged for her lover Vita Sackville-West to take over the fortnightly 'New Novels' talks from Mary Agnes Hamilton, was brazen about the poor BBC rates, doubting that she could 'screw more out of the old misers'.[26] By Vita's standards, the £370 a year it was calculated she could earn (which included the reprinting fee for *The Listener*) may have seemed paltry but it was evidently possible for those who appeared regularly to earn a sizeable sum.[27] In Matheson's defence, she did believe that Vita would be a superb broadcaster. One woman who was identified very early on as having a gift for enchanting and inspiring the listener was Marion Cran.

8.2   MARION CRAN, CELEBRITY GARDENER

Marion Cran, 'the doyen of all listeners who are also amateur gardeners' was one of the BBC's first radio stars.[28] She was one of scores of women invited to Savoy Hill to broadcast as part of the pioneering *Women's Hour* programme yet was the only one to have lasting appeal. Her 'lush and sentimental' talks on gardening made her 'astonishingly popular' and propelled her to national fame.[29] She broadcast her first *Gardening Chat* on 6 August 1923 and fortnightly thereafter. Very possibly her Fleet Street connections (she worked on *Burlington Magazine*) meant she was known

[25] BBC/WAC: Margaret Bondfield Talks:1, Programme Contracts Executive to Bondfield, 21 January 1938.

[26] Hilda Matheson Letters (hereafter HML), 18 January 1929.

[27] HML, 17 February 1929.

[28] *Radio Times*, 6 August 1929.

[29] BBC/WAC: RCONT 1: Marion Cran Talks (hereafter MCT): Creswell to Barnes, 17 September 1937.

to Ella Fitzgerald, the programme's producer. Gardening as a career for women had become a possibility since the late 1880s when the first horticultural colleges were established.[30] Cran, a Fellow of the Royal Horticultural Society, had written two gardening books giving her the requisite expertise to broadcast.[31]

What made Cran distinctive was her warm, informal style.[32] A mature woman in her forties, she had a natural ability to engage with the audience and to speak directly to them. Her scripts were written ungrammatically and conversationally, her language was florid and descriptive. This is immediately evident in the preface to *Garden Talks* a book of her radio broadcasts published in 1925. Claiming to be 'no expert' Cran was instead 'a learner talking to fellow learners', she was a poet, a dreamer, an observer. 'We have ranged the days, from dawn to dusk, and the seasons through and back again and discussed how to make cuttings … how to make paths and little formal town-gardens … and have talked of roses … wild flowers and birds', she effused.[33] Her appeal was not so much the practical as the lyrical. She transported her listeners to lush countryside and beautiful outdoor spaces. Cran cultivated a sizeable and appreciative audience, cementing her celebrity by publishing more than a dozen gardening books.

Cran's talks were still a regular afternoon fixture when Hilda Matheson assumed responsibility for the Talks Department in 1927. Although her timeslot had been on a Saturday (still a workday for most employees, with Sunday the only official day off) as part of the switch to *Household Talks*, Cran was moved to Friday afternoons. Despite her longevity and popularity, Cran's fee remained five guineas. The higher status of evening broadcasts is evident from negotiations she conducted in March 1929 where she successfully raised her fee to the standard ten guineas for two specialist 7.00 pm talks. These were given under the auspices of the National Gardens Guild and involved writing 1,500 words 'in a rather particular technique'.[34] Cran pooh-poohed a request that these talks be published in *The Listener* pointing out how a talk that was 'colloquial, easy, intricate'

---

[30] The Women's Farm and Garden Association was founded in 1899.

[31] Mrs George Cran (1913) *Garden of Ignorance* (London: Herbert Jenkins); (1921) *Garden of Experience* (London: Herbert Jenkins).

[32] Simon Elmes (2012) *Hello Again: Nine Decades of Radio Voices* (London: Random House) pp. 34–6.

[33] Marion Cran (1925) *Garden Talks* (London: Methuen).

[34] MCT, Cran to Sprott, 1 March 1931.

would not make a good magazine article, but similarly, how a 'scholarly, literary' talk would sound bone-dry.[35] This was a moot point. There was frequently tension between Matheson and Richard Lambert, editor of *The Listener*, about the transposing of a talk to the written page.[36] Cran also jested whether, in these evening talks, she would have to become more 'highbrow', perhaps a reference to Matheson's intellectual predilections.

Evidently a friendship developed between Cran and her producer Elise Sprott. Sprott visited 'Coggers', Cran's home in Kent and Cran was in turn invited by Sprott to BBC events such as a 'do' at the Corporation's Sports Ground, Motspur Park.[37] This close association would become problematic for Cran. Her demise on the airwaves coincided with Sprott's transfer from the Talks Department in the summer of 1931. Almost certainly Cran was displaced by C.H. Middleton who broadcast his first gardening talk in May 1931. Without Sprott to champion her, Cran's link with the BBC was severed. Middleton also had a very different style. Both a down-to-earth and captivating speaker, he offered listeners practical advice, particularly on vegetables.[38] With an expansion in suburban gardening, this was seen as more apposite and he quickly became indispensable. Later, from 1933, Vita Sackville-West became the voice of the beauty of gardening, similar territory to Cran's, which made her return impracticable. In 1937, the drama producer Peter Creswell contacted the Talks Department on Cran's behalf (they were old friends) alerting them to Cran's feelings of grief and hurt that, despite her great popularity, she had not been asked back to the BBC since Savoy Hill.[39] The memo was passed on to the producer of women's talks Janet Quigley who contacted Cran. However, her suggestion for a series on 'Gardens I Have Seen' never came to fruition.[40]

Cran's ten years as a broadcaster reveal how someone expert in their field could become a national star. Not everybody liked her whimsical style but her ability to engage the listener made her immensely valuable to the Corporation. Cran spoke predominantly to the female daytime audience; C.H. Middleton's appeal, on the other hand, was such that it embraced men and from the mid-1936 his weekly *In your Garden* slot was moved

[35] MCT, Cran to Sprott, c. April 1931.
[36] Richard Lambert (1940) *Ariel and All his Quality* (London: Gollanz) pp. 112–14.
[37] MCT, 21 June 1929, 12 September 1929, 30 June 1930.
[38] For a discussion on C.H. Middleton see Elmes, *Hello Again*, pp. 61–5; Andrews *Domesticating the Airwaves*, pp. 66–81.
[39] MCT, Creswell to Barnes, 17 September 1937.
[40] MCT, Quigley to Cran, 2 November 1937.

from Friday evening to Sunday afternoon. Cran's relationship with Sprott may also have been significant. She might have protected a woman who had become her friend from being discarded, even though her style was no longer deemed appropriate. Ray Strachey's broadcasting career was also, to some extent, dependent on the whim of her producers but rather than a singular expertise, she built up a reputation of being able to talk with authority on a variety of subjects.

8.3   RAY STRACHEY, DAYTIME ALL-ROUNDER

It is not surprising that the feminist campaigner and writer, Ray Strachey, spoke about politics and citizenship in her BBC talks, but she also spoke about books, travel, careers for girls and even construction work.[41] Her broadcasting career, which spanned 12 years, was launched by Matheson. The two women undoubtedly knew each other as both had been employed as political advisors by Nancy Astor.[42] Matheson first contacted Strachey about the possibility of broadcasting in April 1927 although her first actual talk for the BBC was in December, an evening talk in which she looked back 100 years to 'The Flapper of 1827'.[43] In January 1928 Strachey chaired the 9.15 pm topical debate on 'Need We Envy our Grandchildren?' (Strachey would become a grandmother in 1934) and three months later she gave her first afternoon talk on Southern Italy in the series 'Holidays Abroad'.

A Woman's Commentary, broadcast from January 1929, would bring Strachey to far greater prominence. This was a regular series, part of the new look *Morning Talks*. In October 1928 she and Matheson had lunched together at The Lady Golfers' Club of which Matheson was a member.[44] It was here that the idea of the 'purely experimental … weekly budget of events' of special interest to women was discussed.[45] Women aged over 21 had recently been enfranchised and, with a General Election looming, Matheson was keen to introduce more talks on citizenship. The first *A Woman's Commentary* was scheduled for 9 January. The arrival of the

---

[41] *Radio Times*, 20 December 1927.

[42] Barbara Caine (2005) *Bombay to Bloomsbury: A Biography of the Strachey Family* (Oxford: Oxford University Press) pp. 242, 316.

[43] BBC/WAC:LE(E)1A: Contributors: Mrs Ray Strachey:1a (hereafter RST:1a), Matheson to Strachey, 6 April 1927; *Radio Times*, 20 December 1927.

[44] RST:1a, Matheson to Strachey, 8 October 1928.

[45] RST:1a, Matheson to Strachey, 23 November 1928.

manuscript only the day before (Strachey had been away in Vienna), caused some consternation as the potentially controversial subject matter—on public affairs—meant it had to be checked through by the Director of Programmes, Roger Eckersley. Strachey described spinning it 'out of hot air' and nothing untoward was found.[46]

Strachey, who was paid the standard five guineas for each broadcast, was often self-deprecating about the quality of her talks, referring to them as 'drivelling', 'light and frivolous' and 'very harmless'.[47] Matheson however, was delighted, describing the new feature as 'an excellent one', with Strachey setting the standard high to produce 'an absolutely impartial and yet lively commentary' and one that had found a very appreciative audience.[48] The few surviving scripts (from the 1930s) show them to be both informative and personal. Mrs 'A.B', a housewife from Manchester, was patently pleased. In a letter to *Radio Times* in February 1929, she expressed her thanks to Mrs Oliver Strachey for her recent morning talk. 'To one, at least' she wrote, 'the task of cleaning a kitchen went down a little better whilst listening to the intelligent observations of an intelligent woman'.[49] The success of *A Woman's Commentary* gave Matheson the idea of similar talks by women MPs, paving the way for *The Week in Parliament* which was broadcast from November 1929.

Strachey had been an active member of the National Union of Women's Suffrage Societies (NUWSS) and wrote, *The Cause*, a history of the fight for the vote published in 1928.[50] Millicent Garrett Fawcett, the President of the NUWSS, died the following year and Matheson invited Strachey to broadcast a description of the Memorial Service held at Westminster Abbey.[51] Matheson was always keen to bring elements of culture to her daytime audience and from September 1930, Strachey presented a six-part morning series on 'Reading for Fun' which included talks on romantic books, crime and short stories. This was followed by a second series 'Books about People' in early 1931 which introduced listeners to the joys of biography. There was to be a final series of *A Woman's Commentary* that year after which Strachey's broadcasts become far more sporadic. Once Matheson was no longer in command, it seems the connection with the Talks Department weakened.

[46] RST:1a, Strachey to Matheson, letter wrongly dated 7 December 1927.
[47] RST:1a, March 12 1929, May 21 1929, 27 May 1929.
[48] BBC/WAC: RCONT 1: Mrs Ray Strachey, Talks, Matheson to ?, 3 July 1929.
[49] *Radio Times*, 1 February 1929.
[50] Caine, *Bombay to Bloomsbury*, pp. 310–20.
[51] *Radio Times*, 19 December 1929.

Margery Wace, who assumed responsibility for *Morning Talks* after Matheson's departure, was dismissive of Strachey, informing the new Director of Talks Charles Siepmann that although useful for an occasional talk, she was 'inclined to be rather heavy and monotonous week after week'.[52] Yet Siepmann liked Strachey's ideas and, aware that she had a strong track record with the Talks Department, arranged to meet her.[53] Although her proposed script on 'Family Life' was dismissed, Strachey was invited, in 1934, to broadcast a short series of morning reports about 'Whitehall', a subject she knew well.[54] During 1935, she made two broadcasts: an appeal for Open-Air Nurseries as part of *The Week's Good Cause*, and a short talk on 'The Census and Women's Occupations' for *At Home Today*.[55] It was Janet Quigley who returned her to prominence and the two women worked closely on *Careers for Girls*. As Secretary of the Women's Employment Federation, this was Strachey's mission, although it took almost two years before the six-part series finally came to fruition in April 1939.[56] Strachey also persuaded Quigley of the merits of a talk on housebuilding. She was then in the process of building, by hand, her Mud House at Fernhurst in Sussex which, as an amateur engineer, she had designed herself.[57] 'Building a House' which included descriptions of laying pipes, transporting bricks and fitting windows, was broadcast as a *Teatime Talk* in December 1937.[58]

It was not only the daytime output aimed at women that Strachey took part in. Quigley made use of her for the evening debate series *Men Talking*, as we shall see, and she also put her forward as the presenter of a new series in 1938 about the news.[59] Quigley, however, was absent from the meeting at which this was discussed and the idea was scuppered by Wace who reiterated her view that Strachey was too dull.[60] In the event, it was

---

[52] BBC/WAC:LE(E)1b: Contributors: Mrs Ray Strachey:1b (hereafter RST:1b), Wace to Siepmann, 20 September 1933.

[53] RST:1b, Siepmann to Strachey, 14 December 1933.

[54] BBC/WAC:LE(E)1c: Contributors: Mrs Ray Strachey:1c (hereafter RST:1c), Wace to Strachey, 13 February 1934.

[55] Broadcast on 9 June 1935, 7 February 1935.

[56] RST:1c, Strachey to Quigley, 23 August 1937.

[57] Mary Agnes Hamilton gives a vivid description of the house. Mary Agnes Hamilton (1944) *Remembering my Good Friends* (London: Jonathan Cape) pp. 265–6.

[58] *The Listener*, 12 January 1938.

[59] BBC/WAC:R51/115/1: The Week in Westminster:1a, Wace to Maconachie, 12 April 1938.

[60] BBC/WAC:R51/115/1, Maconachie to Quigley, 21 April 1938.

another seasoned broadcaster, Mary Agnes Hamilton, who presented *Have You Been Following the News?* Strachey was to broadcast once more in 1940, commemorating the twenty-first anniversary of women being granted the vote. Later that year, it was Quigley who informed the Director of Talks, Maconachie, about Strachey's sudden death on 17 July 1940; she had been due to appear on *Calling All Women* the next day.[61] It was Hamilton who broadcast an appreciation of her close friend on the programme the following week.

Ray Strachey was an example of a composite broadcaster with an ability to talk eloquently on an assortment of topics. Although largely confined to the daytime schedules, she occasionally appeared in the evening representing 'the woman's point of view'. The contrary opinions about Strachey's voice and abilities reveal the subjective nature of the relationship between broadcaster and producer. Beatrice Webb was treated very differently, her great renown as a leading social scientist confined her talks to the evenings, her voice and suitability never questioned.

8.4   BEATRICE WEBB, EVENING GRANDEE

'I was in a devil of a funk as I walked along the Embankment to Savoy Hill', Beatrice Webb entrusted to her diary in February 1928.[62] Webb was on her way to deliver her first BBC talk; her reminiscences of the philosopher Herbert Spencer, whom she had known when she was young. Matheson had first approached Webb with the idea for the talk in November 1927, part of a wider series of recollections of notable Britons. Webb had undergone a voice test, a rehearsal and been provided with the leaflet of hints about broadcasting, but she need not have been apprehensive about the actual broadcast; alone in the studio her nerves fell away and she actually enjoyed herself. 'I had hardly any consciousness of being listened to, so private and quiet was the place one was in' she confided, finishing her talk 'with a pleasant sense of successful achievement'. Matheson was clearly thrilled, describing the talk as 'one of the best we have ever had' and that Webb had shown herself to be 'a born broadcaster'.[63] Matheson was always

---

[61] RST:1c, Quigley to Maconachie.

[62] Norman and Jeanne MacKenzie, eds. (2000) *The Diaries of Beatrice Webb* (London: Virago) pp. 464–5. Diary entry, 29 February 1928.

[63] BBC/WAC:910/Mrs Sidney Webb Talks:1 (hereafter BWT:1), Matheson to Webb, 23 March 1928.

on the lookout for erudite talent and saw great potential in Webb, inviting her to do poetry readings. Webb, understandably, was more interested in promoting her own interests and suggested instead a series on studying social facts, a subject that could draw on her 'forty years' experience' in social research and which could introduce listeners to the new discipline of Social Science.[64] With Matheson keen on the idea, Webb drove a hard bargain for her fee, pointing out that the preparation of the 'short tabloid lectures', the rehearsals involved and the adaptation to a new technique were an arduous task for a woman of her age (Webb was almost 71). A payment of £50 was agreed for the four talks.

Described in the BBC's 'Talks and Lectures' brochure as being 'of outstanding importance', the series, *How to Study Social Questions* came to air in March 1929. Webb's views on oral evidence, the use of statistics and the possibility of a 'science of society' were also published in the newly launched, *The Listener*.[65] Webb's future broadcasts would continue to be on topics connected with her work and beliefs. 'Looking Backwards: The World of Politics' broadcast in January 1930 (part of a 17-part series of reminiscences), reflected on changes she had observed in her lifetime, the 'most momentous transformation', she believed, being the demise of government by a 'tiny clique'.[66] 'Taking the Strain off Parliament,' in July 1930, voiced her views on devolution, prompting an editorial in *The Listener* which drew attention to this 'courageous' talk.[67]

Webb's connection with the BBC was not limited to the Talks Department. She also worked closely with the Adult Education section, in particular with Mary Adams and Charles Siepmann. Adams was invited to Passfield Corner, the Hampshire home Beatrice shared with her husband Sidney Webb, now a peer of the realm.[68] (Although efforts had been made to coax Sidney to broadcast, these were always declined.) Adams failed to persuade Beatrice Webb to take part in a series on 'The Census' in 1931, but she did agree to contribute to Siepmann's ambitious venture *The Modern State* in early 1932 agreeing a fee of 50 guineas for her three talks on 'The Diseases of Organised Society'.[69] The manuscript for the first broadcast required Adams' 'blue pencil', with changes sought to its length,

[64] BWT:1, Webb to Lambert, 9 October 1928.
[65] *The Listener*, 13 March 1929.
[66] *The Listener*, 22 January 1930.
[67] *The Listener*, 30 July 1930.
[68] BWT:1, 14 October 1930.
[69] Mrs Sidney Webb Talks: 2 (hereafter BWT:2), Siepmann to Webb, 26 June 1931.

(as a difficult topic it needed to be read slowly and with pauses, so was too long); its use of 'pentasyllabic words', (these needed to be substituted with something simpler); and its casual references to His Majesty (who would probably be listening in).[70] The script for the final talk, to be aired on 21 January, while 'magnificent' was seen to be 'so daring that we have been having conferences over it'.[71] Four corrections were asked for, most notably the removal of a reference to 'mutilation, death and chronic disease' which, Adams agreed, was no stronger than the truth, but which might be 'too strong for some listeners'. Webb was sanguine about the changes, her main annoyance with the series was the way *Radio Times* had advertised her as Lady Passfield rather than Mrs Sidney Webb, her accustomed name.[72]

Russia was to be the subject of Webb's next talk, on 22 September 1932. She and Sidney had spent three months from May to July touring the Soviet Union, which had become popular with left-wing activists. The couple returned full of enthusiasm for the way it was run, their experiences published in 1935 as *Soviet Communism: A New Civilisation?*[73] Webb promised that her BBC talk, which was to be part of the series *Travellers in Europe,* would be objective in tone.[74] Writing in her diary on her return from Broadcasting House she confessed that, although delivered with verve, she had felt 'very unequal to the strain' and had come home with a racing heart.[75] The BBC's reaction to the talk is not recorded but Webb evidently believed it was the reason she was not invited back, commenting to Adams in February 1934 that her broadcasting days were over.[76] Adams was eager that Mrs Webb should return, her suggestion that she speak again on Russia in November 1934 (a talk which did not materialise) perhaps a final chance to use 'the grand old lady'.[77] Webb did agree to contribute to the series *Efficiency and Liberty* in February 1938. The talk was on the Russian system of government and included a question-and-answer session with Henry Wilson Harris, Editor of *The Spectator.*

---

[70] BWT:2, Adams to Webb, 5 January 1932.
[71] BWT:2, Adams to Webb, 19 January 1932.
[72] BWT:2, Webb to Adams, 18 December 1931.
[73] For a discussion on the Webbs in Russia see *Oxford Dictionary of National Biography,* Beatrice Webb, entry 36802 by John Davis.
[74] BWT:2, Webb to Adams, 24 August 1932.
[75] MacKenzie, *The Diaries of Beatrice Webb,* p. 510. Diary entry, 22 September 1932.
[76] BWT:2, Webb to Adams, 16 February 1934.
[77] BWT:2, Adams to Dawnay, 26 November 1935.

This was to be Webb's last broadcast. Now aged 80, her association with the BBC came to a close.

Beatrice Webb was one of a small number of women such as the MP Nancy Astor who made regular appearances in the BBC's interwar evening schedules, their status in public life coupled with their gift at the microphone making them immensely attractive to the BBC. Mrs Edna Thorpe, on the other hand, was unique amongst regular women broadcasters in the interwar BBC, her lack of eminence her selling-point.

8.5   Mrs Edna Thorpe, 'Ordinary', 'Average' Housewife

On 20 March 1935 Edna Thorpe, a London-based housewife, wrote her first letter to the BBC suggesting a short talk.[78] Because of her deep interest in the 'mistress v maid controversy' she divulged, she had gone undercover to investigate the issue for herself, applying for the position of a 'daily help' in a number of houses where she had been able to scrutinise the attitudes of both. Before she sent her observations to the newspapers, Thorpe wondered, might the subject be of interest to the BBC? The letter was passed to Margery Wace who responded with eagerness. A first-hand experience such as this would have been music to her ears and Mrs Thorpe was invited to meet Wace at Broadcasting House and, subsequently, to write a draft script.[79] Impressed by the scope and quality of the writing, as well as Thorpe's pluck, Wace was convinced the talk should be placed in the evening schedules. After much deliberation, in January 1936, in conjunction with a male GP, Thorpe took part in a 6.50 pm discussion on 'The One Maid Problem' for which she was deemed to be a success.[80]

Edna Thorpe would have a broadcasting career that spanned 15 years. Almost all we know about her, however, is what can be gleaned from the hundreds of letters she wrote to the BBC bursting with her effervescent ideas, pursuits and opinions. From these it is possible to ascertain that she was an audacious Yorkshire woman in her early to mid-thirties, she was married to an electrician, she did not have children and, judging by her frequent changes of address, lived in London in rented accommodation.

[78] BBC/WAC: RCONT1: Edna Thorpe Talks:1(hereafter ETT:1), Thorpe to BBC, 20 March 1935.
[79] ETT:1, Wace to Thorpe, 22 March 1935, 3 April 1935.
[80] ETT:1, Wace to Siepmann, 11 June 1935.

The BBC described her as 'lower middle-class'; she portrayed herself as the 'ordinary', 'average' housewife.[81] A fervent listener, Thorpe would turn her lifestyle and varied interests into copy for her talks. BBC producers welcomed this ordinariness and also came to value her sentiments as representative of 'the woman's point of view'. In addition to her constant stream of programme proposals, Thorpe set up informal, and eclectic, listening groups, feeding back sentiments on a wide range of BBC output, a veritable powerhouse of suggestions and ideas.

The series that would become Thorpe's stomping ground was *At Home Today* which Wace had introduced into the morning schedules in September 1934 and which would continue to be championed by her successor, Janet Quigley. Between 1936 and the programme's demise in 1942, Thorpe broadcast talks on topics as varied as the 'White Fish Industry', 'Forming a Reading Club', 'Hints on Springcleaning', 'Notes from a Housewife's Diary' and the woman's angle on the British Industries Fair; the talks she actually gave only a fraction of those she suggested. In the spring of 1936, Thorpe took part in 'The House that Jack Built', a three-part series based on the premise of a housewife taking on the building industry in the form of an architect, a builder and a town planner. Thorpe was felt to be ideal because she was neither professional nor connected with any housing movement, rather, she was 'an intelligent woman who has combined running her own home with finding out how others run theirs'.[82] Two of the individuals tackled by Thorpe were the master builder John Laing and the grandee of town planning George Pepler. To put each programme together the scripts of the two discussants were mailed between them, gradually built up into questions and responses and then rehearsed before broadcast. In the programme with John Laing, for example, Thorpe took him to task on problems of damp, soundproofing, squeaky floor boards, ill-fitting doors as well as the better use of recesses for storage.[83]

As will be discussed shortly, Thorpe was one of a handful of women invited to take part in *Men Talking* and she also participated in two further evening talks. The first was in January 1938 when she was part of a panel discussion in the weekly programme *The Cinema*.[84] The topic

---

[81] For example, ETT:1, Thorpe to Wace, 13 January 1936; Thorpe to Quigley, 4 January 1937; BBC/WAC: RCONT1: Edna Thorpe Talks:2 (hereafter ETT:2), Quigley to Midland Region Director, 5 October 1938.

[82] ETT:2, Luker to Dowler, 16 April 1936. George Luker produced the series.

[83] Broadcast 27 April 1936.

[84] Broadcast 10 January 1938.

was 'The Audience' and Thorpe, 'speaking as a housewife', included the observation that it was one of the 'cheapest ways of getting a change from the daily round' offering the chance 'to be amused or interested in some aspect of our life which seems different from our own'.[85] The second was in *The Poet and the Public* in May 1938. *Radio Times* enthused that she had been brought to the microphone by Humphrey Jennings 'chiefly because she is a typical housewife yet happens to take a great interest in poetry'.[86] Thorpe had first indicated her zest for the subject two months earlier when, on spotting an advanced notice of the series, she suggested her 'intelligent interest in modern poetry' might make her representative of the 'ordinary average reader'.[87] The 'dialogue' was excessively prepared beforehand with Thorpe and Jennings' initial discussion drafted as a script, amended and then twice rehearsed before transmission.[88] After the talk (for which she was paid the standard ten guineas evening fee), Thorpe was pleased to report that she had received many comments about the spontaneity of the broadcast, hinting at her proficiency on air.[89]

Thorpe was exceptional in the interwar years, a regular broadcaster whose expertise was her ordinariness. After 1939, she would continue to be used widely by the BBC, taking part in many of Quigley's wartime programmes for women and later she became a popular speaker on *Woman's Hour*, her final broadcast in April 1951. Thorpe's pre-war appearance on *Men Talking* in October 1937 was specifically to represent 'the woman's point of view'.

8.6   'MEN TALKING' AND 'THE WOMAN'S POINT OF VIEW'

Men Talking was conceived towards the close of 1936 as a series of experimental afternoon discussions aimed predominantly at the unemployed.[90] Two or three speakers 'of the "man-in-the-street" type of intelligence' would discuss, impromptu, without any script, a subject of everyday

---

[85] World Film News No.2, February 1938, quoted in Jeffrey Richards (1984) *The Age of the Dream Palace: Cinema and Society 1930–1939* (London: Routledge) pp. 16, 65.

[86] *Radio Times*, 24 May 1938.

[87] ETT:2, Thorpe to Luker, 23 March 1938.

[88] ETT:2, Luker to Thorpe, 12 April 1938.

[89] ETT:2, Thorpe to Cox, 29 May 1938.

[90] BBC/WAC:R51/118: Debates and Discussions, Rose-Troup to Graves, 16 November 1936.

interest.[91] Topics featured in the first series, broadcast in spring 1937, included the inheritance of wealth, bringing up children, manners and the emphasis of sport in the press. Aimed in particular at discussion groups, the response was far greater than anticipated, with both working-class and middle-class audiences demonstrating approval.[92] As a result it was decided to move the series to the evening schedules in the autumn of 1937.

Before the new series was broadcast, a change of title to 'People Talking' was suggested, since women speakers would now take part.[93] However, the consortium of Talks Assistants who produced the programme, Roger Wilson, George Luker and Janet Quigley, decided that to alter the title would seem 'rather affected' as the emphasis was on the Talking rather than the Men and the original name was retained.[94] There was a strong theme of family life running through the 12 subjects proposed for the ten-week run such as 'The New Freedom in the House', 'The Size of Families' and 'Should Married Women Work outside the Home?' much of which merited 'a woman's point of view'. A panel of 30 potential 'A' list speakers was drawn up which included eight women, amongst whom were Mary Stocks, Olga Collett (who Quigley had recently made the acquaintance of) and Rosalind Johnson, a kitchen maid.[95]

The first *Men Talking* of the autumn series, on 'Parents and Children', caused a riotous response, primarily because of the inclusion of Mrs Winifred Parsons. Luker had met Parsons at a meeting in Cambridge and had thought her 'admirably suitable'.[96] However, the night after the discussion was aired on 7 October, Wilson sent a highly emotive note to Luker which informed him that he had listened to the programme with a group of 'highbrows' who all criticised the programme for being too middle-class, the chief blame for which fell on Parsons.[97] Quigley was also of the view that Parsons had been a disaster; 'she talked too much, she interrupted unscrupulously, she sounded aggressive'.[98] Despite this, Quigley informed Maconachie (the Director of Talks) that both she and

---

[91] BBC/WAC:R51/319: Men Talking (hereafter MT), Maconachie to Graves, 23 April 1937.

[92] MT, Notes on the experience gained running this series January–April 1937.

[93] MT, Maconachie to Graves, 23 April 1937.

[94] MT, Wilson to Maconachie (in consultation with Quigley), 21 June 1937.

[95] MT, Luker to Maconachie (in consultation with Quigley), 13 July 1937.

[96] MT, Luker to Maconachie, 28 September 1937.

[97] MT, Wilson to Luker, 8 October 1937.

[98] MT, Quigley to Maconachie, 8 October 1937.

Luker were in agreement that women should continue in the rest of the series but, in light of the Parsons experience, their role would be reviewed. Firstly, a woman should not be allowed to monopolise the conversation. Secondly, because Parsons was a good example of her type, 'intelligent, vivacious, used to public speaking, eager to express her views' it was this sort of woman who was wrong. In future, programmes should use quieter, non-aggressive women, 'who will only speak when the woman's point of view, as opposed to the man's, is really called for'. Thirdly, women of whatever type needed coaching beforehand, which would include warnings on interrupting. Fourthly, the male speakers must be told not to show more politeness to the women speaker than to each other. 'We hope, by these means' Quigley assured her boss, 'the women speakers will not continue to be a complete liability'. As Luker informed Wilson, 'Miss Quigley is going to experiment with a mouse-like woman instead of the tiger we have been experimenting with in the past'.[99]

Further memos focus on issues of class rather than gender but of the nine remaining programmes in the series, only two included women, Edna Thorpe and Ray Strachey. Strachey's appearance on 18 November was on 'A Woman's Place Is in the Home' in which she defended women's right to work against the novelist Nigel Balchin.[100] For this she received rare praise from John Gloag, who hosted the debates. Admitting that he believed it to be 'an utterly mistaken policy' to include women in the programmes because they 'paralysed the men', he described Strachey as excellent and the exception, someone who illuminated the discussion without dominating it.[101] Thorpe was similarly praised by Quigley for her appearance on 'Family Budgets'.[102] Thorpe's letters to Quigley, however, reveal her exasperation at the lack of women in *Men Talking*, commenting at the close of the series that 'no discussion on any aspect of family life is complete without a woman's angle'.[103]

The heated discussions that surrounded the inclusion of women in *Men Talking* illuminate the contested role of women broadcasters in the interwar years. The success of Strachey and Thorpe was not because they were mouse-like women but because they understood the medium of radio. With women's restricted access to the airwaves, it is self-evident that there

[99] MT, Luker to Wilson, 11 October 1937.
[100] Broadcast on 18 November 1937.
[101] MT, Notes on the Men Talking Series by John Gloag, 28 December 1938.
[102] ETT:1, Quigley to Thorpe, 15 October 1937.
[103] ETT:1, Thorpe to Quigley, 18 December 1937.

would be a far smaller pool of experienced speakers to draw from. In addition, the art of debating would have been taught to, and encouraged in, boys who attended public school, something that would have been far less common for girls. Reflecting on the Parsons experience, Luker was clear that beforehand, she had seemed 'exactly right'. She was accustomed to WEA audiences, she had a pleasing voice ('really' he added), as well as a sense of humour and skill in putting her ideas together. In addition, reports from 50 Listening Groups had all 'singled out for praise Mrs Parson's voice'. The script also showed that what she said was usually sensible.[104]

The issue, then, was purely her behaviour on air. By talking too much, interrupting and appearing aggressive she had not acted in a way that was judged suitable for a woman. She was, after all, only there to represent 'the woman's point of view'. That a woman might be able to contribute to a general discussion was never a consideration. This is evident from the personalised broadcasts that developed in the interwar years, when men (always men) such as Harold Nicolson spoke in broad terms about life, in Nicolson's case the very popular *People and Things*.[105] Women could talk about their expertise, hence Strachey's *Women's Commentary* on political and civil life, but they did not have the authority to talk in general terms, unless it was from a feminine perspective. This may have been linked to the notion that men could represent a national viewpoint, the viewpoint of 'everyman'. And this became even more pertinent when it came to representing the voice of the BBC, as an announcer.

8.7   Women Announcers, the Sheila Borrett Experiment

On 28 July 1933, Sheila Borrett became the first woman announcer on the BBC's National Service. Already known as a radio actress, she had auditioned for the position taking the same test as the men. It was a prestigious job. Wearing full evening dress, BBC announcers were not only expected to introduce programmes but also to deliver football results, the fat-stock prices, SOS messages, weather reports—and the all-important news bulletins. Yet it was more than this; announcers, whether in Britain, the USA, Germany or elsewhere were perceived as the embodiment of

---

[104] MT, Luker to Wilson, 11 October 1937.
[105] Broadcast between 1930 and 1935.

that particular network or station, the 'public image', their voices setting the overall character and tone.[106] At the BBC, as early as 1925 it had been recognised that the announcer built up 'in the public mind a sense of the BBC's collective personality'.[107] This was why, from 1924, BBC announcers were anonymous, it was not their personality but the Company/ Corporation's that they were projecting. However, unlike her male counterparts, it was impossible for Mrs Giles Borrett to remain unknown.[108]

Provincial and regional BBC stations had, from the early 1920s, used female announcers to introduce their women's talks and *Children's Hour*. In the summer of 1930, there had been alarm in the press when it appeared that a woman had been appointed to a permanent position in the Midland Region.[109] A BBC official quickly moved to quell fears, explaining that Miss Gladys Ward was being used merely as temporary holiday relief.[110] The London *Evening News* was reassured. Having expressed horror about 'a Hobb's century being announced in a pleasant soprano or the details of a heavy-weight fight related by a girlish voice!', the paper was glad to report that men who listened 'may breathe again, secure in the knowledge that all depressions, all scores and all that Mr Snowden [the Chancellor of the Exchequer] may do, will be announced henceforth in fine resounding baritones'. The *Evening News* would have been aghast to learn that many European-based stations such as Radio Luxembourg, Radio Normandie and Radio Budapest regularly used female announcers. A feature in *Radio Pictorial*, entitled 'Eva at the Mike', profiled nine such women in countries that included Lithuania, Switzerland, Denmark, Finland, Italy, Spain and Poland.[111] German radio also employed female announcers who, although rare on the general output, were the predominant voice on women's programmes.[112] They were also commonplace in the USA with women widely accepted as announcers and presenters on the prolific daytime schedules aimed at the female audience.[113]

---

[106] Michele Hilmes (1997) *Radio Voices: American Broadcasting, 1922–1952* (Minneapolis: University of Minnesota Press) pp. 58–9; Kate Lacey (1996) *Feminine Frequencies: Gender, German Radio, and the Public Sphere, 1923–1945* (Michigan: University of Michigan Press) p. 199.

[107] Briggs, *Birth of Broadcasting*, p. 292.

[108] In 1925 announcers were required to wear dinner jackets.

[109] *Evening World; Daily News and Chronicle*, 21 July 1930.

[110] *Evening News*, 25 July 1933.

[111] *Radio Pictorial*, 10 January 1936.

[112] Lacey, *Feminine Frequencies*, p. 196.

[113] Hilmes, *Radio Voices*, pp. 141–4.

In her 1933 book *Broadcasting* Hilda Matheson pondered the question
as to why women announcers, who were favoured in Latin countries and
Central Europe, were unwelcome in the UK. She thought it likely that the
'immense importance attached by the British to sport' meant that there
was a sense that 'no woman could read football and cricket results with
the peculiar conviction which a male voice alone would convey to them'.
She did, however, express incredulity that this should be put forward as an
objection as 'many male announcers themselves neither play nor are interested in these manly games'.[114] Notwithstanding Matheson's viewpoint,
Radio Times previewed the arrival of the BBC's first woman announcer
with foreboding, foreseeing 'panic among the horsehair armchairs, retired
colonels muttering darkly over their muffins, consternation in the bow-
windows of the historic clubs'.[115] Borrett's first job was to introduce a teatime concert from London's Hotel Metropole. Reviewing the debut of the
'woman with golden voice', *News Chronicle* declared that she had 'good,
clear vocalisation, correctly pitched, pleasing in its cadency' although it was
pointed out that she was being paid £500 a year just to say a few words.[116]
The paper also made much of her 15-month-old son who, they claimed,
was listening in. As a wife and mother, Mrs Giles Borrett's appearance at
the microphone transgressed, in many people's minds, the acceptable role
of a woman, in particular a married woman. On the other hand, she encapsulated modernity, a further reason why the press were so fascinated by her.

The incongruity of Borrett's position was compounded when, on 21
August 1933, she read the six o'clock news bulletin for the first time.
Europe had again led the way in its promotion of women newsreaders. In
1926 the Stuttgart-based station Surag had appointed a woman to read
the headlines, weather reports and financial information. Norag, a pioneering Hamburg station, followed suit shortly after and was soon boast-
ing two women announcers. In 1932, Gertrude Van Eyersen became the
first woman announcer on national German radio (the Berlin Funkstunde)
although her appointment was met with contempt in the press.[117] This
was due not only to questions about a woman's authority to speak for the
nation but also because, in hard economic times, she was seen to be taking

---

[114] Matheson, *Broadcasting*, p. 56.

[115] *Radio Times*, 4 August 1933.

[116] *News Chronicle*, 29 July 1933, quoted in Anne Karpf (2007) *The Human Voice: The
Story of a Remarkable Talent* (London: Bloomsbury) pp. 158–9.

[117] Lacey, *Feminine Frequencies*. Lacey points out that it was still rare for women to fill
these roles, pp. 200–1.

a job from a man. Authority was the underlying reason why women in the USA were debarred as announcers from the 'proper' business of radio which was 'men's' concerns such as news, politics and sport.[118] Elsie Janis, NBC's first woman announcer in 1935, was quickly moved away from reading news bulletins because of listener complaints.[119]

Sheila Borrett was to suffer a similar fate. In October 1933, after only three months in the job, she was axed. In its statement to the press the BBC explained that it was because they had received more than 10,000 letters of complaint with more than 90 per cent of detractors being female.[120] In a personal article 'Why I Came and Why I Went' published in *Radio Pictorial* the following year Borrett wondered whether it was the pre-publicity that had done for her.[121] There had been mass anticipation of her taking on the role which was seen as a daring experiment. Perhaps if she had started unobtrusively, she mused, she might have been given the chance to grow into it. She also pointed out that nearly 90 per cent of the letters of appreciation that she had received had also been from women. Characterising the English as a conservative people who hated innovations, she also raised the issue of discrimination: was it 'prejudice against women playing a speaking part in the affairs of the world?' she wondered. Although no longer an announcer, Borrett continued to feature widely on the BBC appearing in a host of dramas and as a reader of poetry and prose, seemingly with no objections. Her final appearance at the microphone, in December 1937, was a weekly serialisation of Jane Austen's *Northanger Abbey*.

The vitriol that marked Borrett's arrival as a radio announcer was in sharp contrast to the delight that accompanied the entrance of Jasmine Bligh and Elizabeth Cowell as the first female Television Hostess-Announcers in November 1936. Selected from hundreds of hopefuls, the two young women were expected to exude poise and glamour. A *Daily Mirror* report on their first press conference described Bligh as 'tall, statuesque, really beautiful in the dignified Edwardian manner'. Cowell, conversely was 'slight, quick, with a lively face which one would call "chic"'. The former had blue eyes, the latter brown and they both had 'pleasant voices, easy manners and were discretely dressed in sober black and white'.[122] A BBC survey into viewers' opinions carried out in 1939, on whether television

---

[118] Hilmes, *Radio Voices*, pp. 141–4.
[119] Karpf, *The Human Voice*, p. 159.
[120] *The Daily Mail*, 6 April 1934; *Daily Express*, 7 April 1934.
[121] *Radio Pictorial*, 9 March 1934.
[122] Quoted in Gordon Ross (1961) *Television Jubilee, The Story of 25 Years of BBC Television* (London: WH Allen) p. 32.

announcers should be male or female, showed that while 44 per cent were indifferent, the overwhelming majority of the rest preferred women.[123] Cowell and Bligh's success was because of their appropriateness. They did not assume to speak for the nation in a position of authority rather they were there to introduce viewers to the new visual wonder of television, their beauty, style and composure central to the role.

Although Borrett's tenure as a radio announcer was brief, BBC management were theoretically prepared to give women another chance. In June 1934, the Board of Governors resolved that there was to be no ban on the employment of women as announcers, actively encouraging further experimentation in the Regions.[124] In January 1939 it was further agreed that a 'woman announcer should be experimentally employed when a convenient opportunity occurs'.[125] During the Second World War women would be widely used to announce programmes, though not to read the news. The continuing antipathy towards women newsreaders is encapsulated in a June 1939 *Radio Times* article by the journalist and novelist Irene Stiles. An impassioned plea for the 'wider use of the feminine voice in broadcasting', it emphatically did not include women reading the news. 'Even the most ardent feminists' Stiles claimed, 'would find it difficult to produce a woman's voice that could deal calmly with so many world-stirring events'. It was the 'quiet unruffled' tones of the male announcers that proved soothing and reassuring. 'I, for one, would not have that otherwise' she declared.[126] Irene Stiles scepticism about women reading the news did not extend to women as commentators. In particular, she exuded praise for Mrs Olga Collett who, she believed, offered 'the perfect illustration of how adequately a woman's resourceful tongue and all-embracing eye can deal with brilliant functions and picturesque events'.

8.8   WOMEN COMMENTATORS, THE TRIUMPH OF OLGA COLLETT

Olga Collett 'Britain's Ace Radio Talker' was unique, the only woman in the interwar years to work regularly for the BBC's Outside Broadcast Department as a commentator.[127] 'Eye witness' accounts had been an

[123] Briggs, *The Golden Age of Wireless*, p. 622.
[124] BBC/WAC:R1/3/1: Board of Governors, Minutes, 27 June 1934.
[125] BBC/WAC:R1/7/1: Board of Governors, Minutes, 25 January 1939.
[126] *Radio Times*, 23 June 1939 'More Women Commentators' by Irene Stiles.
[127] *Radio Pictorial*, 16 June 1939.

important radio fixture from 1924, the skill of commentating very different to that of announcing. Rather than reading a carefully prepared script in a studio environment, to be a commentator involved relaying quick-witted and highly descriptive observations from live sporting fixtures and national events. Howard Marshall, Tommy Woodroffe, John Snagge and Teddy Wakeham were amongst the popular voices who, in the interwar years, regularly broadcast from football matches, boat races, public commemorations and royal proceedings.

Collett's route to commentating was via Talks, she broadcast her first on 'Political Canvassing' in February 1937 as part of Janet Quigley's *At Home Today* series.[128] Emboldened by the experience, in March 1937 Collett wrote to Seymour Joli de Lotbiniere, the Director of Outside Broadcasts requesting an audition 'with a view to my name being placed upon your panel of women commentators'.[129] Her letter arrived at an apposite time. The sportswoman and journalist, Marjorie Pollard (after a number of successful talks and many approaches to the BBC) had been invited to give an eye witness account of the England v Australia Women's Cricket Test Match in June 1937, the first time this had been attempted.[130] Pollard, who played both cricket and hockey at a national level, was an obvious choice. She was editor of the magazines *Hockey Field* and *Women's Cricket* and she also contributed sporting commentaries to *The Times, The Observer* and *The Evening Standard* revealing an appetite for a woman's eye view on female sports within the male-dominated world of sports journalism.[131] Through Pollard, the BBC was hoping to tap into this active interest.

Pollard was an undoubted expert who was known to be an able broadcaster, even if her 'vigorous' voice did have a tendency to 'squeak at moments of excitement'.[132] Lotbiniere was at this time also scouting for a woman commentator for a broader sphere of work and a number of

---

[128] Broadcast on 25 February 1937. Quigley and Collett developed a close working relationship and Collett broadcast many talks.

[129] BBC/WAC: OB Commentary: Mrs Olga Collett (hereafter OCC), Collett to Lotbiniere, 30 March 1937 .

[130] BBC/WAC: RCONT1: Marjorie Pollard Talks: 1, Programme Contracts to Pollard, 30 April 1937.

[131] *Oxford Dictionary of National Biography*, Marjorie Anne Pollard, entry 65061, by Judith Wilson. See also Adrian Bingham (2004) *Gender, Modernity, and the Popular Press in Inter-War Britain* (Oxford: Clarendon Press) pp. 69–74.

[132] Marjorie Pollard Talks:1, Wilson to Midland Regional Director, 27 April 1936.

hopefuls had been put through his 'Roof Test' which involved being taken by lift to the top of Broadcasting House and being asked to describe the surroundings.[133] Olga Collett's roof top audition took place on 16 April 1937.[134] Recalling the occasion in a 1983 radio interview, she divulged how, when she asked Lotbiniere how long she was to speak for he said, 'well, everyone else we've tried has dried up within two or three minutes—and at the end of 11 minutes he came and begged me to stop!'[135]

Collett claimed that her gift of description and ability to hold an audience were acquired in the village halls and public meetings she addressed when her brother, the Liberal MP Kingsley Griffith, was first contesting his parliamentary seat.[136] She was an authoritative figure with a full-time job as Female Staff Administrator for ICI, a position similar to that of the BBC's Miss Freeman. As she revealed in Quigley's series *Other Women's Lives*, it was the death of her husband, Squadron Leader Stanley Collett, in an RAF flying display in 1934 that propelled her to find work.[137] Her position as a widow meant that her appearances on the BBC were less problematic than those of Borrett who, as a married woman, was both stigmatised for shunning her wifely duties and accused of taking jobs away from men. The fact that Collett had an established career, however, impacted on her ability to work for the BBC; any events that she covered had to be slotted into her leisure time.

Collett's successful audition had taken place just weeks before George VI's Coronation for which she was eager to be considered as a commentator. As she pointed out in a letter to Lotbiniere, a woman would have 'a more noticing eye for details' as well as bringing relief during the long broadcast. It was also timely, she believed, for a woman to be considered.[138] Lotbiniere's response, while thanking her for putting 'the woman's case', pointed out that the BBC were not keen to risk such an experiment with Coronation Day.[139] Instead, Collett's first commentary was on 17 June 1937 when she was paid eight guineas to describe

[133] BBC/WAC:R30/428/1: Commentators, 22 January 1937. Seven people were trialled on this day including two women.
[134] OCC, Lotbiniere to Collett, 14 April 1937.
[135] BBC Sound archive, 41963, Olga Collett interviewed by John Lane, 26 July 1983.
[136] This was in a closely contested by-election in March 1928 for the seat of Middlesbrough West.
[137] 'Other Women's Lives', 29 May 1937.
[138] OCC, Collett to Lotbiniere, 21 April 1937.
[139] OCC, Lotbiniere to Collett, 22 April 1937.

'the arrival of Their Majesties' at Royal Ascot.[140] This may well have been the occasion Lotbiniere had had in mind when he recruited Collett. Her credentials for such a broadcast were impeccable. Already known to the royal family through her late husband (the Prince of Wales was honorary Air Commodore of Stanley Collett's RAF squadron) her father-in-law was Sir Charles Henry Collett, Lord Mayor of London in 1933.[141]

Collett always insisted that she was a commentator, *not* a woman's commentator. She was not employed to talk about fashions rather it was the scene, the movement, the action and the colour that she described.[142] Nevertheless, her Ascot commentary was called 'frivolous' by the *Daily Herald,* causing her to vent her spleen that this was based on 'the usual assumption that women are only capable of describing dresses'. In a forthright letter to Lotbiniere she expressed resentment that it was a BBC employee Andrew Stewart (Edinburgh Programme Director) who had made this claim and a second claim that no woman could come into line with men on a serious or symbolic occasion. Had any woman ever been given the opportunity of tackling such an event, Collett wanted to know.[143] Collett was to be given the opportunity in November 1937 when she depicted the scene at the Guildhall for the Lord Mayor's Banquet, an assignment she repeated the following year. In its preview, *Radio Times* made much of her family connections and her new positioning in the Guildhall Library, where she was to describe the assembled procession as the dignitaries walked in to dine.[144] Collett also attended a display by the Women's League of Health and Beauty at Wembley in June 1939 and the European Figure-skating Championships in January 1939 where Cecilia College, the reigning British Champion, won gold. This particular broadcast was deemed to be of such high quality that it was used for many years by the BBC's Staff Training Department as an example of a perfect commentary.[145]

It was an event in March 1939, however, that brought Collett overnight fame. As part of the State Visit by the French President to London, a Gala Concert had been arranged at the Royal Opera House, in the company of the King and Queen. Collett was to provide part of the commentary, her

[140] BBC/WAC:R1/73/7: Board of Governors: DG's Reports and Papers, 14 July 1937.
[141] 'Other Women's Lives'
[142] Olga Collett interview.
[143] OCC, Collett to Lotbiniere, 25 June 1937.
[144] *Radio Times,* 28 October 1938.
[145] BBC Sound Archive 10293, 'What is Good Radio', 22 November 1946.

stint extended when she was obliged to stand in for a male colleague who had been taken ill. With the Royal party then delayed by more than 20 minute, she found herself on air far beyond her allotted time. Her highly personalised descriptions of the outfits, the textures, the sounds, 'a word picture of brilliance', made headline news the following day, *The Daily Mail*, *The Star* and the *Evening Standard* all applauded her tenacity and style.[146] It was only belatedly that BBC executives arranged to raise the fee for her 'marathon broadcast' to 12 guineas, aware that 'Howard Marshall would have certainly got 18–20 guineas for a similar effort'. It was also agreed that in future her fee should be at least ten guineas for events of national importance.[147] There were to be no such occasions; with the outbreak of war Collett's career as a commentator came to an end. She did continue to broadcast wartime talks for Quigley and, with Quigley as Editor, appeared on *Woman's Hour* several times in the 1950s. Collett would also be a panellist on early editions of *Twenty Questions*.

8.9   CONCLUSION

Collett's achievements as a commentator in many ways exemplified what it took to be a successful woman broadcaster on the interwar BBC. She built strong relationships with producers who nurtured and promoted her; her links to public life and her association with royalty gave her an authority to speak; her sphere, though she insisted it was not fashion, was still appropriate for a woman, describing the spectacle of state functions and feminine events.

To be a broadcaster was to enter the nation's homes. She or he personified the BBC's public service ethos to inform, educate and, occasionally, to entertain. The live nature of the medium meant that, once the impromptu pioneering days were past, scripts were carefully checked and, where possible, rehearsed. Matheson's professionalisation of broadcasting, with the introduction of guidelines for the way a talk should be written and delivered, clarified the intimate nature of the spoken word and marked out a special role for those who were able to captivate and engage.

The women considered in this chapter represent different facets of British broadcasting in the interwar years. Marion Cran was one of the earliest radio celebrities, her passion for gardens and her way of inspiring

[146] *Evening Standard*, 23 March 1939.
[147] OCC, Outside Broadcast Executive to Talks Booking Executive, 11 May 1939.

the listener by evoking the beauty and poetry of her craft saw her gain an appreciative following and kept her in work at the BBC for eight years. Cran's arena was the female daytime audience, which was also Ray Strachey primary domain. Strachey's strength was her ability to turn her hand to a breadth of topics whether books, careers or citizenship, subjects to which, as a clever and accomplished older woman, she was able to bring an informed point of view. But both Cran and Strachey, it appears, were vulnerable to the whims of the Talks Department. Once the producer who championed them had left, they fell out of favour, in Cran's case permanently. The eminence of Beatrice Webb protected her from this, it was her status as a national figure which had brought her to the BBC, the quality of her scriptwriting and the attractiveness of her voice then ensuring that she remained in demand. Webb only ever spoke in the evening schedules, something occasionally achieved by Strachey and by Edna Thorpe, in her capacity as representing the woman's point of view. Thorpe's broadcasting career was predicated on her status as a housewife. Her capacity to turn her life experiences into engaging talks made her invaluable to a BBC that was aware of this missing voice.

The very different ways in which the Corporation dealt with Sheila Borrett and Olga Collett highlight the fine line that was trodden when speaking on behalf of the BBC. Borrett was never castigated by her employers; it was the volume of listener complaint that prompted her removal. Collett, as an outside broadcaster, was highly praised for her commentary, the public events she described did not overstep the line of authority that Borrett was seen to have breached. Women's voices and their authority on air have continued to be the subject of fierce debate. In the interwar years, thousands of women spoke before the microphone many of whom were widely appreciated and a small number garnered fame. But the contested role of women in public life meant that they were never used to the same extent as men, their right to talk for the nation only occasionally sanctioned. Like the institutional BBC, women were largely denied access to the most highly prized roles.

CHAPTER 9

Conclusion

In September 1939, *Weldon's Ladies Journal* published an article headlined 'Women of the BBC'.[1] It took the same format as most previous press stories, with a strapline that claimed, 'Behind the scenes at Broadcasting House are many busy women, each in her own way contributing to your enjoyment when you switch on your radio'. Ruth Maschwitz, who wrote the article, had been taken on a tour of the kitchens by the restaurant supervisor Mrs Dubarry, discussed the day-to-day work of the Drama Department with Mary Hope Allen and Barbara Burnham, stopped by the Reference Library, where Miss Milnes was constantly answering queries on the phone and marvelled with Mrs Webbsmith at the majestic flower display she had arranged in the foyer. Maschwitz ended her article with the declaration 'away with the idea that radio is a man-run concern … each department of this vast organisation in one way or another owes something to the energy and inspiration of the women of the BBC'.

Maschwitz's account of the Corporation's female employees on the brink of the Second World War captures the essence of the BBC in the 1920s and 30s. It was a place where women in all capacities from house staff to administrators to creatives could play an important role. Indeed Hilda Matheson, as Director of Talks and Mary Somerville, as Director of School Broadcasting, were pivotal to its development. The BBC was a remarkable place for women to work, an organisation where conditions of service were good and where there were possibilities to excel and achieve.

[1] *Weldon's Ladies Journal,* September 1939.

© The Editor(s) (if applicable) and The Author(s) 2016
K. Murphy, *Behind the Wireless,*
DOI 10.1057/978-1-137-49173-2_9

251

In an era when many professional women continued to experience overt and entrenched discrimination, the BBC's ethos of equality and its acceptance of the intrinsic value of women had a positive impact on what was broadcast on air.

The *Weldon's Ladies Journal* article also testifies to the way in which BBC women were frequently utilised by the popular press and women's magazines as a symbol of modernity. And it was the modernity of the BBC that set it apart from many British workplaces in the interwar years. It was a post-First World War industry in a brand new area of employment. Broadcasting in Britain was instituted by the BBC. It had no blueprint when it started and it developed in an ad hoc fashion creating a host of new careers in its wake. Women, like men, were part of this pioneering evolution 'having a go at any kind of job'.[2] At the start, many of these jobs were ungendered which allowed able and ambitious women to appropriate them, the exponential growth of the Corporation then establishing them in significant roles. Women such as Florence Milnes in the Reference Library, Kathleen Lines in the Photographic Section and Elise Sprott as Woman's Press Representative had joined the young Company in its first few years. By 1939 they were all earning in excess of £500 a year in positions where they were held in high esteem.

Sprott and Milnes had joined as waged staff and this was another feature that set the BBC apart from other areas of women's employment in the 1920s and 30s. Whereas much office-based work was 'dead-end', at the BBC the incremental grading system meant that those who remained with the Corporation could rise into salaried posts where status and conditions of service were heightened. Yet one of the reasons why significant numbers of salaried women began their BBC careers as waged was because, for them, recruitment directly to the 'senior' ranks was far harder. This was partly due to a deeply ingrained lack of self-confidence. As Winifred Holtby passionately declared, young women were used to being treated as second-best to their brothers and seeing only men 'at the top of all trees', so it was not surprising if they had doubt in their own capacities.[3] It was also because there was an acceptance that women would start this way. Ray Strachey noted that women had mostly entered new spheres of work at the

---

[2] *Prospero*, December 1968, recollections of Winifred Boustead, filing clerk, Magnet House in early 1923.

[3] Winifred Holtby (1934) *Women and a Changing Civilisation* (London: Lane and Bodley Head) pp.101–4.

bottom 'often bringing abilities too good for the tasks they have been set to do'.[4] This was certainly the case with Elizabeth Barker. Looking back at the start of her BBC career in 1934 she recalled how in those days 'it was so difficult for a girl to get an even modestly interesting job – even with a respectable degree and a good deal of foreign travel behind her – that I didn't grumble, or at least not for four years'.[5]

Once in the salaried grades, women had to negotiate the 'public school' character of the BBC. It was awash with men who had found their way to its doors through 'influence'. 'Old School Tie-ism run amok' as one contemporary commentator declared.[6] Although some women arrived at the BBC through friends and family connections, this was almost always to the secretarial and clerical grades. There is, however, little evidence of overt hostility towards women. This was because they were rarely, if ever, in direct competition with men. They carved out their own specialisms and areas of work and so long as they were not seen as a threat, men were chivalrous towards them and often enthusiastic about them. Whether it was praise for Agnes Mills' improvements in the Registry, satisfaction at Mary Candler's enhancements to the way copyright was approached or celebration at Ursula Eason's much heralded decision to broadcast the Christmas edition of Northern Ireland *Children's Hour* from a ward in the Belfast Hospital for Sick Children, salaried women were widely acknowledged to be performing good work.[7] For those who reached top jobs, there was an appreciation of their cleverness and intellect. Hilda Matheson and Mary Somerville were undoubtedly sharp and highly intelligent. They were also pleasant to work with, and they were different, which added an extra frisson. They were also extremely diligent. They did not view their jobs as a stepping-stone to something higher, rather they wanted to perform to the best of their ability within that role.

The BBC was in the vanguard of British employers who introduced forward-looking practices in the interwar years. Whereas in teaching and the Civil Service attempts to further women's advancement in terms of equal opportunities were fiercely opposed, in the BBC they were largely embraced, at least in principle. And while women were often disadvantaged in terms

[4] Strachey, *Our Freedom*, p.143.
[5] *Prospero*, December 1974.
[6] Garry Allighan (1938) *Sir John Reith* (London: Stanley Paul) p.235.
[7] *Radio Pictorial*, 18 December 1936; BBC/WAC:L1/306/1: Agnes Mills Staff File:1, Confidential Report February 1930; BBC/WAC:R13/296: Gramophone Department, Women Clerical, Fletcher to Rose Troup, 17 February 1938.

of promotion and pay, discrepancies, if realised, could be broached with management. Unlike teachers and civil servants, BBC women were not faced with indissoluble prejudicial policies. This was most evident in the application of the Corporation's marriage bar which, in the case of the salaried, was adroitly circumvented.

The spectre of marriage shaped attitudes towards all young female employees in the interwar years. It was a reality that, for most, work was a temporary affair. This understanding led the BBC to develop the notion of 'two classes of women': those who intended to remain with the Corporation permanently and 'those whose mind is not here but in their homes', a notion that became the driving force behind the 1932 bar. While this was not specifically directed at the waged, it was here that the impact was most keenly felt. To identify the 'exceptional' woman was the over-riding function of the ill-fated Marriage Tribunal. It made good economic sense to retain a valuable employee and the benefit to the BBC of a dedicated, hardworking married woman was never in doubt. The BBC flew in the face of the widely held orthodoxy that matrimony and a career were impossible. In 1939, its three highest paid female employees were married, two of them mothers, a scenario unimaginable in most workplaces at this time.

The marriage bar symbolised an underlying tension that dogged the BBC throughout the interwar years, the clash between the progressive and the conventional. The BBC enjoyed its image as a modern, enlightened organisation, its acceptance of women an important element. On the other hand, it was desirous to be viewed as part of the establishment, as a revered national institution. This would become progressively more obvious in the 1930s, with the Corporation increasingly likened to the Civil Service. This straddling of the modern and the traditional was apparent in many aspects of women's presence at the BBC, not least in how they dressed. The bespectacled librarian in her tweeds offered a very different image to the glamorous shorthand typist in her fur. Employment practices also spanned the old and the new. The long-standing custom of segregation and defined female roles, under the control of the Women's Staff Administrator, was, for most of the BBC's waged secretarial and clerical staff, common practice. Salaried women, conversely, worked largely as equals with men, in the same grades, with the same salary bands and to the same managers.

The tension between the modern and the traditional was also evident in the BBC's output aimed at women. Those with the responsibility for the broadcasts deliberated on how best they should inform, educate and entertain. Because of the lower status of daytime talks, an element of experimentation was possible and new formats were trialled. Margery Wace's series *How I Keep House*, for instance, was the first to use only working-class voices. Neither Ella Fitzgerald nor Elise Sprott were intellectuals, their more populist approach echoed newspaper fare. Hilda Matheson began the process of professionalisation which was extended by Oxbridge-educated Margery Wace and Janet Quigley. All were driven, however, by a deep desire to improve women's lives whether this was through enlightening them on the role of a magistrate, diverting them with tales of a Japanese schoolgirl or advising them on how to help a stammering child.

In a three-page article 'Broadcasting for Women' published in *Woman's Magazine* in October 1939 (but obviously penned before the outbreak of war) Elise Sprott pondered how to best address the topic. Should it be about the important women who had spoken during the years of broadcasting or the women who planned and produced programmes not only for women but for listeners as a whole?[8] In fact, she elucidated on both, picking out Dame Henrietta Barnett, Amy Johnson and Olga Collett for special commendation before the microphone and Janet Quigley, Olive Shapley and Barbara Burnham for their production prowess behind. Without doubt, women programme makers brought women's voices to the BBC, from Ray Strachey commentating on careers for girls and on the joys of building a house to Mrs Emerson describing the week she spent living with a miners' family in France. Matheson, in particular, enticed to Savoy Hill a raft of female grandees, although they were always far fewer women broadcasters of evening fare. The authority to represent the voice of the nation, as an announcer, also remained problematic. The experiment with Sheila Borrett was tried and quickly dropped. Olga Collett, however, was accepted as the voice of the BBC in her commentaries showing that, provided she did not stray into the male bastion of news, a woman could be admired.

No woman at the interwar BBC ever breached the bastion of the Control Board. Reith, while accepting of able women, was not intrinsically a modern man. His comfort zone was an all-male environment. That the highest echelon of the BBC should be in the hands of men was self-evident to him, it never entered his head that it should be otherwise. But whereas in 1926, Hilda Matheson could be head-hunted and

[8] *Woman's Magazine*, October 1939.

given the job of Director of Talks and, in 1931, Mary Somerville confirmed as Director of School Broadcasting, after 1933, when Isa Benzie became Foreign Director, no other woman in the interwar BBC reached a director-level post. By the mid-1930s, ensconced at Broadcasting House and with an ever-expanding and ponderous hierarchy, the institutionalisation and professionalism of the Corporation made it far harder for women to attain a top job. Eminent academics, civil servants and former politicians, like Professor John Coatman (appointed Head of News, 1934), Sir Stephen Tallents (appointed Controller of Public Relations, 1935) or Sir Richard Maconachie (appointed Director of Talks, 1936) accepted senior executive BBC posts while elsewhere male graduates jostled for departmental supremacy.

Although it might be harder to get to the top, women who were ambitious and talented continued to thrive at the BBC. Dozens of those who joined the BBC in the interwar years would go on to have impressive careers into the 1940s and 50s: Clare Lawson Dick, Mary Lewis, Mary Candler, Anna Instone, Marie Slocombe, Ursula Eason and Elizabeth Barker, to name but a few. And what of Elise Sprott? She retired as Head of Section, Lecture and Women's Interests in 1945. In 1939 she was at the peak of her BBC career, doing a job that she loved, that drew on her passion for public speaking, for travel, for sharing her enthusiasm for the BBC. She was not a highbrow or a high-flyer. She was a doughty, hardworking woman who grasped opportunities that came her way. A broadcaster, an administrator, a producer and a communicator, she epitomised the explosion of possibilities that were available to women in the early BBC. Sprott was neither typical nor atypical. It was the combination of the modern and the conventional and the acceptance of young and old, graduate and non-graduate, married and unmarried, glamorous and plain, that made the BBC in the 1920s and 30s such an extraordinary place for women to work.

EPILOGUE: A BRIEF ENCOUNTER WITH 90 FURTHER YEARS

Volumes could be written about the women at the BBC post-1939; these few pages offer just a glimpse of significant changes and individuals; the seeds of further research.

The Second World War, like the First, was a time of great fluctuation for women in the UK, particularly in relation to work.[1] At the BBC, as thousands of men were called up, women took over essential jobs, most strikingly as engineers. An Engineering Training School opened in May 1941, the first seven female trainees arriving the following month. By the end of the war more than 800 women would pass through its doors, being deployed first on maintenance and programme work and later at transmission stations. Sir Noel Ashbridge, the BBC's Chief Engineer, declared in *The BBC Year Book* for 1943, that 'the experiment of recruiting women and training them for technical work has been an undoubted success'.[2] The BBC more than doubled in size during the war creating an urgent need for more support staff. However, the introduction of conscription for single women (aged between 20 and 30) in December 1941 meant there was a severe shortage of 'suitable' secretarial recruits. As a result a Secretarial Training School was established in February 1942, its aim, 'to convert a mere typist into an instructed and enthusiastic employee

---

[1] There are many books on women and work in the Second World War, see particularly Penny Summerfield (1989) *Women Workers in the Second World War: Production and Patriarchy in Conflict* (London: Routledge).

[2] *BBC Year Book,* 1943, p.88.

© The Editor(s) (if applicable) and The Author(s) 2016   257
K. Murphy, *Behind the Wireless,*
DOI 10.1057/978-1-137-49173-2

of the Corporation'.[3] Another new and vital area of work was the BBC's Monitoring Service, based at Caversham near Reading. Hundreds of female linguists arrived, many of whom were mothers, and because of the need for 24 hour working, a full-time crèche was opened in 1943 at nearby Sonning Manor.

The expansion of the BBC's overseas services also created new opportunities for women. Una Marson, a Jamaican, was the first black woman to join the BBC staff in March 1941 (she had briefly worked before the war on a contract basis as a researcher for the TV programme *Picture Book*).[4] Marson was employed as a Programme Assistant on *Calling the West Indies* but also, in 1943, developed the long-running literary series *Caribbean Voices* (although Marson would leave the BBC in 1945). Cecelia Reeves, who had been one of Isa Benzie's Assistants in the Foreign Department, played a key role in the French Service. As Senior Talks Assistant she coordinated the team of French broadcasters from London, including Jacques Duchesne. Reeves was one of a number of women who rose to senior positions in the External/World Service after the war. She became the BBC's Paris Representative in 1947, a position she held for 20 years.

The Television Service was shut down at the outbreak of war. Elizabeth Cowell, a former TV Hostess, joined the growing band of women who were now accepted as radio announcers. Margery Anderson, Jean Metcalfe, Margaret Hubble, Joan Griffiths and Mary Malcolm, for instance, began their BBC careers at this time. Women's programmes continued to be an important element of the broadcast schedules, most of which were produced by Janet Quigley. Series such as *The Kitchen Front, Wise Housekeeping, Calling the Factory Front* and *Your Health in Wartime* provided essential information as well as boosting morale.[5] Audrey Russell was spotted as a broadcasting talent (she had been working as an auxiliary firewoman) and was recruited to the BBC in June 1942.[6] In 1944, she became the Corporation's only female accredited war correspondent, covering events in mainland Europe. After the war she gained acclaim as an outside commentator, particularly celebrated for her coverage of the Royal Wedding

---

[3] BBC/WAC:R13/241: Secretarial Training Centre, Burlton to Controller (Administration), 6 October 1941.

[4] See Delia Jarrett-Macauley (1998) *The Life of Una Marson, 1905–65* (Manchester: Manchester University Press).

[5] Sian Nicholas (1996) *The Echo of War: Home Front Propaganda and the Wartime BBC, 1939–1945* (Manchester: Manchester University Press) pp.70–107.

[6] Audrey Russell (1984) *A Certain Voice* (Bolton: Ross Anderson).

in 1947. Russell had been prepared to accept unequal pay during the war. However, in peacetime, she complained about her lower earnings and her sidelining into 'women's' issues, a situation rectified when she threatened to resign.[7]

After the war, BBC Monitoring was scaled back, the crèche closed and no further female engineers were recruited. Women engineers who wanted to remain, although no longer employed at transmitters, were retained for studio work but with limited options for promotion because 'they generally did not possess the necessary technical qualifications'. The Secretarial Training School, however, was bolstered under its new head, Marian Scott. She held the position until 1967 when the School was merged into General Staff Training. Women's programming was also strengthened. *Woman's Hour* was first broadcast in October 1946, its mix of domesticity, citizenship and escapism reminiscent of the BBC's pre-war women's fare.[8] In 1950, Janet Quigley (who had left the BBC on her marriage in 1945) returned as Editor, with Olive Shapley as presenter, two of a sequence of high-powered women who would shape the programme both behind and before the microphone into the present day.[9]

Shapley was to be one of the presenters of women's programmes on television. Mary Adams, now a Senior Television Producer, devised *Designed for Women*, in 1947 which was followed by other series such as *For the Housewife* and *Leisure and Pleasure*.[10] Doreen Stephens, who joined BBC television in October 1953 as Editor, Women's Programmes, was the first woman to be directly recruited to a senior post from outside the Corporation since Hilda Matheson.[11] In 1964 the Women's Unit was merged with children's programming into a Family Programmes Department; however, by 1967 the 'women's' element had been dropped. Children's programming had by then been greatly expanded. First with the introduction of *Listen with*

[7] The Oral History of the BBC: Audrey Russell interview, (no date).

[8] Kristin Skoog (2014) 'Striving for Editorial Autonomy and Internal Recognition: BBC Woman's Hour' in Maggie Andrews and Sallie McNamara eds. (2014) *Women and the Media: Feminism and Femininity in Britain, 1900 to the Present* (London: Routledge) pp.99–112.

[9] Jenni Murray (2006) *Woman's Hour: Celebrating Sixty Years of Women's Lives* (London: John Murray).

[10] Mary Irwin (2014) 'Women's Viewpoint: Representing and Constructing Femininity in Early 1950s Television for Women' in Andrews and McNamara, *Women and the Media*, pp.113–26.

[11] Mary Irwin (2011) 'What Women Want on Television: Doreen Stephens and BBC Television Programmes for Women, 1953–64', *Westminster Papers*, Vol. 8, issue 3.

Mother (the brainchild of Mary Somerville) broadcast on the Light Service from 1950 and second by the creation of a Children's Programmes department in Television (initiated by Adams) under the headship of Freda Lingstrom in 1951. Children's television would largely continue to be the domain of women, headed by the likes of Monica Sims, Anna Home and Lorraine Heggessey.

Grace Wyndham Goldie is the best known of all BBC women of the 1950s and 1960s. She moved from Talks Producer, Radio (a post she had been recruited to in 1944) to the position of Talks Producer, Television in October 1948, where her influence would be immense.[12] Amongst her many successes were the pioneering of General Election coverage, the revamping of *Panorama*, the development of *Tonight*, the nurturing of *Man Alive* and the launch of *That Was the Week That Was*. She also championed the BBC careers of, amongst others, Michael Peacock, Richard Dimbleby, Donald Baverstock, Huw Wheldon and Alasdair Milne. Although there were female members of her team, she was not a great advocate for her sex either on air or within the BBC. She was dismissive of female voices, which she felt lacked authority, although she did later concede that she might have done more to promote women.[13] Goldie retired in 1965 as Head of Talks and Current Affairs, Television. Mary Adams, who became Head of TV Talks in 1948, was initially Goldie's boss. Adams was then promoted in 1953 to Assistant to the Controller of TV, a post she held until her retirement in 1958. Mary Somerville also took on a new role becoming Assistant Controller of Talks in 1947 and ultimately Controller of Talks in 1950. She retired in 1955. Janet Quigley and Isa Benzie also assumed important roles, Quigley as Chief Assistant and later Assistant Head of Talks. The two women worked on the development of the *Today* programme which was launched on the Home Service in October 1957. Isa Benzie was its first producer.[14]

Sir Ian Jacobs, interviewed on *Woman's Hour* in 1959 (following his retirement as Director General) was asked which single word he would

---

[12] See, for example, Grace Wyndham Goldie (1977) *Facing the Nation: Television and Politics, 1936–76* (London: The Bodley Head); Charlotte Higgins (2015) *This New Noise: The Extraordinary Birth and Troubled Life of the BBC* (London: Guardian Books) pp.70–90.

[13] *The Sunday Times*, 27 April 1969.

[14] Paul Donovan (1997) *All our Todays: Forty Years of Radio 4's 'Today' Programme* (London: Jonathan Cape) pp.8–14.

choose to characterise his seven years at the BBC. His reply, 'Hag ridden!'[15] By the 1960s, the perception of a BBC teeming with high-flying women could no longer be the subject of jest. Those who had joined the Corporation in the interwar years and risen to positions of authority had largely all left. In 1962, Thelma Cazelet Keir, recently retired as a BBC governor, wrote a letter to *The Times* headed 'Senior Posts at the BBC: Why are there so few women?' They held only four out of 150 top jobs.[16] This lack of women at the top was picked up again in 1968 when Peggy Jay, of the BBC's General Advisory Committee, requested statistical information about the number of women in senior positions. Management assured her that with regard to selection, where there was no inherent advantage of sex, men and women were given equal consideration.[17]

There were significant developments for women in the 1960s. Joanna Spicer (who had arrived at the BBC in 1941) was promoted to Assistant Controller, Planning in 1963 and later, in 1969, Assistant Controller, Television Development, making her by far the most senior woman in the BBC at this time.[18] Although hers was a pivotal strategic role it has been suggested that she was not given the full title 'Controller' because she was female. Verity Lambert was the young producer given the task of developing a new programme, *Dr Who* first transmitted in 1963, the theme music created in the BBC Radiophonic Workshop by Delia Derbyshire. Joan Marsden was employed as a Floor Manager in television, the first and only female to hold this position between 1960 and 1975. Another woman isolated in a male enclave was Yvonne Littlewood, a Light Entertainment producer who, from 1964, worked on programmes such as the *Eurovision Song Contest* and the *Val Doonican Show*. Mary Edmond similarly breached a male stronghold of the BBC in 1964 when she was appointed a Duty Editor in the Newsroom. A further advance for women was the extension of the General Trainee Scheme, introduced in 1954 to fast-track outstanding male graduates. From 1960 this was open to young women, although the numbers recruited remained very low.

---

[15] Leonard Miall (1994) *Inside the BBC: British Broadcasting Characters* (London: Weidenfeld & Nicolson) p.173.

[16] *The Times*, 15 March, 1962.

[17] BBC General Advisory Council, meeting 24 January 1968, Arkell (Director of Administration) to Jay, 29 February 1968.

[18] Miall, *Inside the BBC*, pp.171–6.

The General Trainee Scheme was one of the areas of concern pinpointed in *Women in Top Jobs* published in 1971.[19] The study of the Corporation, carried out by PEP (Political and Economic Planning), confirmed what had by now become obvious to many: the BBC was no longer a progressive institution when it came to the employment of women. They held only 58 out of 1,095 top graded jobs with just one woman (Joanna Spicer) graded 'A'. Amongst the many reasons identified for women's poor progress were lack of career development; the appointment boards system; the paucity of female graduates applying for the General Trainee Scheme; the closure of certain parts of the BBC to women such as engineering and technical work and the scarcity of women in areas such as Current Affairs, Sports, Outside Broadcasts and Light Entertainment. Women, it was found, tended to be clustered in sections where they had traditionally done well, such as television and radio production, children's programmes, schools, further education and make-up. This meant they became typecast with less mobility. While women's personal attitudes were also seen as contributory, the largest stumbling block was the difficulties inherent in juggling family life with BBC work, particularly raising children. The study concluded 'the BBC ought not to assume that it has equal opportunities for women, but should examine what it means by equal opportunities. It cannot ignore the long-term consequences of losing or not recruiting high calibre women'.[20]

Women in Top Jobs prompted the Corporation to carry out its own internal survey of female staff. When 'Women in the BBC' was presented to the Board of Management in April 1973 it caused shockwaves. The report clarified the scarcity of women in senior posts, just 5 per cent, with miniscule numbers at the very top. In addition it provided ample evidence of widespread misogyny, as Jean Seaton vibrantly depicts in her account of women in the BBC in the 1970s and 80s.[21] Senior managers in Engineering, for example, gave reasons for the unacceptability of women that included their unsuitability to heavy physical work and hazards such as bad weather or working alone, marital problems that might result from men and women working together on remote sites, the fact that they were uneconomic because they resigned for family reasons and their inability to

[19] Michael Fogarty, A.J. Allen, Isobel Allen, Patricia Walters (1971) *Women in Top Jobs: Four Studies in Achievement* (London: Allen and Unwin) pp.157–222.

[20] *Women in Top Jobs*, pp.219–20.

[21] Jean Seaton (2015) *Pinkoes and Traitors: The BBC and the Nation 1974–1987* (London: Profile) pp.207–31.

cope with shift patterns as they aged. One manager in television operations was blunt, 'the majority of men would deep down resent women coming into this area and having equal opportunity. Such a move would threaten male superiority'. Explanations as to why so few secretaries moved into higher positions included views that ranged from being regarded 'as a piece of inanimate office furniture rather than a member of a team', that they were 'destined to be wives and mothers and therefore not suitable for further training and help' and that 'a boss with a good secretary doesn't always encourage her to seek promotion'.

The Board of Management were quick to agree a number of recommendations, particularly that no vacancies should be 'for men only' and that women's potential should be positively encouraged and developed. They also concurred that ways should be found to make it easier for women to combine work and domestic commitments. The BBC's first crèche campaign, organised by women staff, had got underway the previous year and in 1974 an experimental nursery was opened at Pebble Mill in Birmingham.[22] However, it closed after a year, the first BBC nursery finally opening in 1990. An external campaign was that of 'Women in Media' who, in 1971, began pressurising the BBC to allow women to read the news. Nan Winton had briefly done so on television in 1960, but was dropped because, like Sheila Borrett in 1934, she was perceived to lack authority. The bastion was finally breached in 1974 when Sheila Tracey became the first regular female newsreader on Radio Four followed by Angela Rippon who, from 1975, presented BBC One's *Nine O'clock News*. In 1979, the authors of *Women in Top Jobs* returned to the BBC to see what progress had been made. They found, however, that although the percentage of women employed in senior grades had risen to around 7 per cent, there were even fewer women in top jobs. 'Years of benign neglect' had resulted in a talent gap.[23]

One of those who had assumed a top job was Clare Lawson Dick who, just ahead of her retirement in 1975, briefly became Controller of Radio Four. In 1978, the position went to Monica Sims. Sims, promoted to Director of Programmes, Radio in 1983, resigned in 1984. However, her association with the Corporation did not end. She was asked to carry out

---

[22] See Suzanne Franks (2011) 'Attitudes to Women in the BBC in the 1970s—Not So Much a Glass Ceiling as One of Reinforced Concrete', *Westminster Papers*, Vol 8, issue 2.

[23] Michael Fogarty, Isabel Allen and Patricia Walters (1981) *Women in Top Jobs, 1968–1979* (London: Heinemann Educational Books) p.161.

a new survey into the position of women at the BBC. The Sims Report 'Women in BBC Management' was published in 1985 and revealed that, while 38 per cent of the monthly paid staff were women, they held only six out of the top 175 posts, less than 4 per cent. Not all was negative, however. Sims noted that there were increasing numbers of women broadcasters, not just newsreaders, but reporters such as Kate Adie and Frances Coverdale and presenters that included Margaret Jay on *Panorama*, Joan Bakewell on *Newsnight*, Esther Rantzen on *That's Life* and Judith Hann on *Tomorrow's World*. There were also high-profile women on radio; Sue MacGregor, Rosemary Harthill, Libby Purves, Gloria Hunniford, Anne Nightingale and Margaret Howard amongst them.

Sims interviewed many female managers about their work, her key findings echoing much of what had been learned before. There was, for example, reluctance amongst women to push themselves forward for promotion; having children meant that many missed out at a crucial point in their careers and the 'men's club' atmosphere at the top was seen to be alienating. One of her recommendations was the appointment of a Women's Employment Officer, along the lines of a successful scheme at Thames TV. She also suggested better career guidance; a review of Appointment Boards policy for senior posts and an increase in the numbers of women on management training courses. Part-time work, job sharing and flexible working should also be encouraged Sims advised. The Board of Managers accepted all recommendations and, in August 1986, the BBC's first Equal Opportunities Officer was appointed. Initiatives soon came thick and fast: the setting up of a training fund, primarily for women-only courses; the creation of a job-share register and training for managers in fair selection. In 1987, Production Secretaries in Radio went on strike, demanding better pay and regrading to Production Assistants, in line with Television. They won.

In 1990, for the first time, the BBC appointed a woman to its Management Board, Margaret Salmon, the new Director of Personnel. Salmon had been recruited from the Burton Group and was, exceptionally, a woman engaged from outside. All other women who held top posts at this time (and in the years before) had risen through the Corporation's ranks. During the nineties, more executive women would be recruited directly into the BBC including Liz Forgan who arrived in 1993 as Managing Director, Radio (from Channel Four), and Carolyn Fairbairn (enticed from the Downing Street Policy Unit), as Director of Strategy at BBC Worldwide in 1997. In October 1990, the Corporation set its first targets for women. At that time, women made up 10 per cent of top managers and 20 per cent of senior and middle management ranks. The targets were 30 per cent women in top management and 40 per cent in senior and middle management by 1996. Although not achieved by the original target date, in 2002, 38 per cent of senior management posts were held by women who now made up almost half the staff.

In March 1991, BBC Television hosted a conference 'Spot the Difference' with an audience of 250 included women broadcasters, journalists and politicians.[24] John Birt, then Deputy Director General, made the opening speech in which he declared that each time he was confronted by a roomful of men at a meeting he was 'struck by the fact that half of those men are standing in the way of talented women'. When 'Opportunity 2000' was launched in October 1991 the BBC, a founder member, committed itself to a range of equal opportunities measures. However, in 1994 an internal survey revealed that women staff were paid on average 25 per cent less than men and in 1996, following a restructure, the number of women on the Board of Management dropped from four to two.[25] During the 1990s, many women who had made their careers with the BBC did reach the top. Jane Drabble, who joined the BBC as a studio manager in 1968, became Assistant to the Controller of BBC One in 1991. Jenny Abramsky, who joined as a studio manager in 1969, rose to be Director of Radio and Music in 1998. Patricia Hodgson, recruited as a producer in Education in 1970 was appointed Director of Policy and Planning in 1993. Three women who joined as news trainees also gained significant posts. Caroline Millington, one of the first trainees in 1970, became Controller of Multimedia Development in 1997. Lorraine Heggessey became Head of Children's Television in 1997 and the first woman controller of BBC One in 2001 while Jana Bennett reached the top job of Director of Television in 2002. Both were news trainees in 1979.

Since 2000, the BBC has continued to take part in a range of initiatives to improve the position of women, including launching a Diversity Strategy in 2011 (in response to the Equality Act 2010) and signing up to 'Project 28-40' in 2014, to increase the proportion of women in technological and engineering roles. However, the number of women in senior management positions has stubbornly remained at around 38 per cent

[24] 'Spot the Difference' conference on the future of women in British television, 14 March 1991, held at Television Centre.
[25] *The Independent*, 17 January 1994.

and women continue to be poorly represented at Executive level. The highest ratio of women to men on the BBC's Executive Board was in 2006, when four out of the nine board members were women. In 2015, the ratio was four out of 13. These have been rocky times for the BBC which was pilloried for age discrimination against women and for its lack of women in authoritative roles. Miriam O'Reilly successfully won her case against the BBC in 2011, after being dropped from her presenter role on BBC One's *Countryfile*. In 2012, Caroline Criado-Perez founded 'The Women's Room' in response to the shunning of women experts by Radio Four's *Today* programme. 'Sound Women', launched as a pressure group for women in radio in 2011, published research in 2014 which revealed women were still seriously under-represented in radio presenting roles.

The BBC, like many large institutions, continues to wrestle with issues of equality, yet women are still attracted to work there. My own experience of 24 years with the Corporation (1987–2011) was, on a personal level, hugely satisfying. I loved my job at *Woman's Hour* where, like my spirited foremothers of the 1920s and 30s, I was able to bring to women's programming my own passions and expertise. Yes, I struggled to juggle family and work as I raised three children. I groaned under endless management restructuring, painful budget and staffing cuts and the frequent crises in which the Corporation became embroiled. But despite the frustrations and limitations, as Lilian Taylor prophesised way back in 1923, the thrill and excitement largely remained. There are many areas of the BBC where women continue to be unrepresented, especially at the very top. Rona Fairhead became the first female chairman of the BBC when she was appointed to head the BBC Trust (the continuation of the BBC Governing body) in 2014. It is intriguing to speculate when a woman will finally take on the mantle of John Reith and be offered the position of 'DG'.

APPENDIX 1

GRADES AND WAGE BANDS: WEEKLY PAID STAFF

Table A.1 Grades for waged women secretarial/clerical staff: January 1927

Grades	Wage rate	Designation
E:	£3.10s to £4.15s	Secretaries and Assistants
F:	£2.10s to £3.15s	Shorthand Typists and Junior Assistants
G:	£2.5s to £3.15s	Copying Typists, Filing Assistants, Telephonists
H:	£1 to £2	Juniors

Source: R1/63/1: Board of Governors. DG's Reports and Papers, January–December 1927. Outline of Organisation, January 1927

Table A.2 Grades for waged women secretarial/clerical staff: April 1937

Grade	Maximum	Standard increase	Appointment
AxW	£6	5s	Senior Secretaries and Clerks
A1W	£5	5s	Senior Secretaries and Clerks
A2W	£4.10s	5s	Mostly promotions from BW on confidential work
BW	£4.10s	5s	Secretaries to Departmental Heads and some Clerks
C1W	£4	5s	'Isolated' Shorthand typists and some Registry Clerks
C2W	£3.15s	5s	Shorthand Typists, Registry Clerks and Telephonists
C3W	£3.10s	5s	High grade Copying Typists, Stencil Typists, Multigraph and Roneo Operators
DW	£3	2/6d	Routine Registry Clerks, Copying Typists and Junior Duplicating Machine Operators
EW	£2	2/6d	Office girls of Duplicating Section

Source: R49/227/1: Staff Policy: Grades and Salaries, Grades 'D' and Weekly Paid Staff

© The Editor(s) (if applicable) and The Author(s) 2016   267
K. Murphy, *Behind the Wireless*,
DOI 10.1057/978-1-137-49173-2

APPENDIX 2

GRADES AND SALARY BANDS: SALARIED STAFF

Table A.3 Grades for salaried staff, January 1927

Grade	Salary	Job
A	£800 upwards	Heads of Departments, Programme Sectional Chiefs Head Office, Regional Station Directors
B	£500 to 1,000	Senior Assistants at Head Office, Sectional Chiefs, Main Station Directors
C	£500 to 600	More responsible Assistants at Head Office, Relay Station Directors and Senior Assistants at Main stations
D	£200 to 400	Junior Assistants, Head Office and provinces

Source: BBC/WAC:R1/63/1: Board of Governors. DG's Reports and Papers, January–December 1927. Outline of Organisation, January 1927

Table A.4 Grades for salaried staff, 1935

Grade	Roof	Increment	Women	Men	Job
A	£1,000 up	£100	1	58	Heads of Departments, Regional, Directors, Chief Editors
B	£1,000	£50	1	49	Regional Musical Directors, Programme Directors, Regional Executives
B1	£800	£50	8	101	Senior Assistants

(continued)

© The Editor(s) (if applicable) and The Author(s) 2016
K. Murphy, *Behind the Wireless*,
DOI 10.1057/978-1-137-49173-2

Table A.4   (continued)

Grade	Roof	Increment	Women	Men	Job
C	£600	£25	28	191	Assistants
D	£400	£20	36	120	Junior Assistants
E	£300	£10	12	18	Bookkeepers and Clerical Staff

Source: These figures are derived from two sources: Salary Information Files and BBC/WAC:R49/231/1: Staff Policy: Grades and Salaries: Monthly (Except Grade 'D'), 1927–1939. Document prepared for the Ullswater Committee, undated but c. 1935

APPENDIX 3

WOMEN WHO EARNED £500 OR MORE PER ANNUM (1939)

Name	Position, April 1939	Start date/earnings	Salary, April 1939
Vera M. Hills	In-Charge, Duplicating	17.3.24/£3	£500
Miss L.G. Maddick	Assistant, Photographs	22.10.34/£312	£500
Gwen Williams	Pianist and Coach, Music	18.7.37/£450	£500
Mary Allan	Television Make-Up and Wardrobe Manager	9.2.37/£380	£520
Evelyn Gibbs	Assistant, Schools	1.10.36/£400	£520
Florence Milnes	In-Charge, Library	26.1.25/£3.10s	£525
Miss A.M. Playle	Assistant to Editor, The Listener	19.1.25/£3	£530
Christine Orr	Children's Hour Organiser, Edinburgh	19.1.37/£500	£550
Cecil Dixon	Accompanist, Music	7.1.23/£150	£600
Elizabeth Jenkin	Children's Hour Executive	4.7.27/£250	£600
Margaret Mackenzie	Press Officer, Bristol	22.7.26/£350	£600
Miss A.M.P. Mills	In-Charge, Registry	12.8.27/£275	£600
Elise Sprott	Assistant, Press (Women's Interests)	29.1.25/£3.15s	£600
Mary Hope Allen	Assistant, Features and Drama	14.6.27 /£3.10s	£620
Janet Quigley	Assistant, Talks	17.2.30/260	£620
Dr Edith McQueen	Assistant, Schools	19.9.30/£375	£645

(continued)

© The Editor(s) (if applicable) and The Author(s) 2016
K. Murphy, *Behind the Wireless*,
DOI 10.1057/978-1-137-49173-2

(continued)

Name	Position, April 1939	Start date/earnings	Salary, April 1939
Doris Arnold	Accompanist & Arranger, Variety	9.2.26/£2.5s	£660
Ella Fitzgerald	Assistant Editor, World Radio	7.4.23/£208	£680
Gweneth Freeman	Women's Staff Administrator	27.4.31/£325	£720
Kathleen Lines	In-Charge, Photographs	7.4.24/£200	£720
Margery Wace	Empire Talks Organiser	8.9.30/£250	£750
Barbara Burnham	Features and Drama Producer	1.4.36/£550	£800
Mary Adams	Producer, Television	5.5.30/£650	£900
Mary Somerville	Director, School Broadcasting	13.7.25/£240	£1,500
Left staff:			
Olive Schill	Assistant, Manchester	17.9.28/£260 (Resigned 31.5.35)	£500
Jo Stanley	Secretary to DG and Governors	2.1.28/£3.10s (Resigned 1.4.38)	£500
Janet Adam-Smith	Assistant Editor, The Listener	5.5.30/£3.10s (Resigned 10.4.35)	£650
Elizabeth Nash	Secretary to DG and Governors	29.9.24/£3.10s (Resigned 1.9.36)	£600
Isa Benzie	Foreign Director	12.12.29/£3 (Resigned 3.1.38)	£900
Hilda Matheson	Talks Director	13.9.26/£600 (Resigned 3.3.32)	£1050

Source: From Salary Information Files: as at 1 April 1939

BIBLIOGRAPHY

BOOKS

Alexander, S. (1995) *Becoming a Woman and Other Essays in 19th and 20th Century Feminist History* (New York: New York University Press)

Allighan, G. (1938) *Sir John Reith* (London: Stanley Paul)

Anderson, G., ed. (1988) *The White Blouse Revolution: Female Office Workers since 1870* (Manchester: Manchester University Press)

Andrews, M. (1997) *The Acceptable Face of Feminism: The Women's Institute as a Social Movement* (London: Lawrence and Wishart)

Andrews, M. (2012) *Domesticating the Airwaves: Broadcasting, Domesticity and Femininity* (London: Continuum)

Andrews, M., and McNamara, S., eds. (2014) *Women and the Media: Feminism and Femininity in Britain, 1900 to the Present* (London: Routledge)

Annan, N. (1985) *Our Age: The Generation that Made Post-War Britain* (London: Harper Collins)

Anthony, S. (1932) *Women's Place in Industry and Home* (London: Routledge)

Avery, T. (2006) *Radio Modernism: Literature, Ethics, and the BBC, 1922–1938* (Aldershot: Ashgate)

Bailey, M. (2009) 'The Angel in the Ether: Early Radio and the Construction of the Household' in *Narrating Media History*, ed. Michael Bailey (London: Routledge)

Banks, O. (1993) *The Politics of British Feminism, 1918–1970* (Aldershot: Edward Elgar)

Beauchamp, J. (1937) *Women Who Work* (London: Lawrence and Wishart)

Beddoe, D. (1989) *Back to Home and Duty: Women between the Wars 1918–1939* (London: Pandora)

Benet, K. (1972) *Secretary* (London: Sidgwick and Jackson)

© The Editor(s) (if applicable) and The Author(s) 2016
K. Murphy, *Behind the Wireless*,
DOI 10.1057/978-1-137-49173-2

Berry, P., and Bishop, A., eds. (1985) *Testament of a Generation: The Journalism of Vera Brittain and Winifred Holtby* (London: Virago)

Beveridge, W. (1960) *The London School of Economics and Its Problems 1919–1937* (London: Allen and Unwin)

Bingham, A. (2004) *Gender, Modernity, and the Popular Press in Inter-War Britain* (Oxford: Clarendon Press)

Biscoe, V. (1932) *300 Careers for Women* (London: Lovat Dickson)

Boyle, A. (1972) *Only the Wind Will Listen: Reith of the BBC* (London: Hutchinson)

Bradley, H. (1989) *Men's Work, Women's Work: A Sociological History of the Sexual Division of Labour in Employment* (Cambridge: Polity Press)

Braybon, G., and Summerfield, P. (1987) *Out of the Cage: Women's Experiences of Two World Wars* (London: Pandora)

Briggs, A. (1961) *The Birth of Broadcasting: The History of Broadcasting in the United Kingdom* Vol. 1 (London: Oxford University Press)

Briggs, A. (1965) *The Golden Age of Wireless: The History of Broadcasting in the United Kingdom*, Vol. 2 (London: Oxford University Press)

Briggs, A. (1979) *The War of Words, The History of Broadcasting in the United Kingdom*, Vol. 3 (London: Oxford University Press)

Briggs, A. (1979) *Governing the BBC* (London: BBC)

Briggs, A. (1985) *The BBC: The First Fifty Years* (Oxford: Oxford University Press)

Brittain, V. (1928) *Women's Work in Modern England* (London: Noel Douglas)

Burrows, A. (1924) *The Story of Broadcasting* (London: Cassell)

Cairns, J.A.R. (1928) *Careers for Girls* (London: Hutchinson)

Caine, B. (2005) *Bombay to Bloomsbury: A Biography of the Strachey Family* (Oxford: Oxford University Press)

Carney, M. (1999) *Stoker: The Biography of Hilda Matheson OBE, 1888–1940* (Llangynog: Michael Carney)

Central Employment Bureau for Women and Careers Association (1931) *Careers and Professional Training: A Guide to Professions and Occupations for Educated Women and Girls* (London: The Women's Employment Publishing Company)

Chignell, H. (2011) *Public Issue Radio: Talks, News and Current Affairs in the Twentieth Century* (Basingstoke: Palgrave Macmillan)

Cohn, S. (1985) *The Process of Occupational Sex-Typing. The Feminisation of Clerical Labour in Great Britain* (Philadelphia: Temple University Press)

Cole, M., ed. (1936) *The Road to Success: 20 Essays on the Choice of Career for Women* (London: Methuen)

Cole, M. (1938) *Marriage: Past and Present* (London: Dent and Son)

Cowman, K., and Jackson, L., eds. (2005) *Women and Work Culture in Britain c.1850-1950* (Aldershot: Ashgate)

Cran, M. (1925) *Garden Talks* (London: Methuen)

Davidoff, L., and Westover, B., eds. (1986) *Our Work, Our Lives, Our Words: Women's History and Women's Work* (London Macmillan Education)

Delap, L. (2011) *Knowing Their Place: Domestic Service in Twentieth Century Britain* (Oxford: Oxford University Press)

Donovan, P. (1997) *All Our Today's: Forty Years of Radio 4's 'Today' Programme* (London: Jonathan Cape)

Dyhouse, C. (1995) *No Distinction of Sex? Women in British Universities, 1870–1939* (London: UCL Press)

Dyhouse, C. (2010) *Glamour: Women, History, Feminism* (London: Zed Books)

Evans, D. (1934) *Women and the Civil Service* (London: Pitman)

Eckersley, P. (1942) *The Power Behind the Microphone* (London: The Scientific Book Club)

Eckersley, R. (1946) *The BBC And All That* (London: Sampson Low, Marston)

Elmes, S. (2012) *Hello Again: Nine Decades of Radio Voices* (London: Random House)

Fogarty. M., Allen, A.J., Allen, I., Walters, P. (1971) *Women in Top Jobs: Four Studies in Achievement* (London: Allen and Unwin)

Fogarty. M., Allen, I., Walters, P. (1981) *Women in Top Jobs, 1968–1979* (London: Heinemann Educational Books)

Gardiner, J. (2010) *The Thirties: An Intimate History* (London: Harper Press)

Gielgud, V. (1947) *Years of the Locust* (London: Nicholson and Watson)

Giles, J. (2004) *The Parlour and the Suburb: Domestic Identities, Class, Femininity and Modernity* (Oxford: Berg)

Glendinning, V. (1983) *Vita: The Life of Vita Sackville-West* (London: Penguin)

Glucksmann, M. (1990) *Women Assemble: Women Workers and the New Industries in Inter-War Britain* (London: Routledge)

Goatman, W. (1938) *By-Ways of the BBC* (London: P.S. King)

Gorham, M. (1948) *Sound and Fury* (London: Percival Marshall)

Graves, P. (1994) *Labour Women: Women in British Working Class Politics 1918–1939* (Cambridge: Cambridge University Press)

Grevatt, W. (1988) *BBC Children's Hour: A Celebration of Those Magical Years* (Lewes: The Book Guild).

Grieve, M. (1964) *Millions Made My Story* (London: Gollancz)

Grisewood, F. (1959) *My Story of the BBC* (London: Odhams)

Hakim, C. (1996) *Key Issues in Women's Work: Female Heterogeneity and the Polarisation of Women's Employment* (London: Athlone)

Hamilton, M.A. (1941) *Women at Work: A Brief Introduction to Trade Unionism for Women* (London: Routledge)

Hamilton, M.A. (1944) *Remembering My Good Friends* (London: Jonathan Cape)

Head, A. (1939) *It Could Never Have Happened* (Kingswood: The Windmill Press)

Hearn, J. (1992) *Men in the Public Eye: The Construction and Deconstruction of Public Men and Public Patriarchies* (London: Routledge)

Hendy, D. (2007) *Life on Air: A History of Radio Four* (Oxford: Oxford University Press)

Hendy, D. (2013) *Public Service Broadcasting* (Basingstoke: Palgrave Macmillan)

Hennessy, B. (2005) *The Emergence of Broadcasting in Britain* (Lympstone: Southerleigh)

Hennessy, E. (1992) *A Domestic History of the Bank of England, 1930–1960* (Cambridge: Cambridge University Press)

Higgins, C. (2015) *This New Noise: The Extraordinary Birth and Troubled Life of the BBC* (London: Guardian Books)

Hilmes, M. (1997) *Radio Voices: American Broadcasting, 1922–1952* (Minneapolis: University of Minnesota Press)

Hines, M. (2008) *The Story of Broadcasting House: Home of the BBC* (London: Merrell)

Holden, K. (2007) *The Shadow of Marriage: Singleness in England, 1914–60* (Manchester: Manchester University Press)

Holloway, G. (2005) *Women and Work in Britain since 1840* (London: Routledge)

Holtby, W. (1934) *Women and a Changing Civilisation* (London: Lane and Bodley Head)

Horwood, C. (2005) *Keeping up Appearances: Fashion and Class between the Wars* (Stroud: Sutton)

Hughes, D.W. (1936) *Careers for Our Daughters* (London: A&C Black)

Hunter, F. (2012) *Hacks and Dons. Teaching at the London Journalism School 1919–1939: Its Origin, Development and Influence* (Colchester: Kultura Press)

Jarrett-Macauley, D. (1998) *The Life of Una Marson, 1905–65* (Manchester: Manchester University Press)

Jennings, H., and Gill, W. (1939) *Broadcasting in Everyday Life: A Survey of the Social Effects of the Coming of Broadcasting* (London: BBC)

John, A.V. (2013) *Turning the Tide: The Life of Lady Rhondda* (Cardigan: Parthian)

Kanter, R.M. (1977) *Men and Women of the Corporation* (New York: Basic Books)

Karpf, A. (1988) *Doctoring the Media: The Reporting of Health and Medicine* (London: Routledge)

Karpf, A. (2007) *The Human Voice: The Story of a Remarkable Talent* (London: Bloomsbury)

Klingender, F.E. (1935) *The Condition of Clerical Labour in England* (London: Martin Lawrence)

Lacey, K. (1996) *Feminine Frequencies: Gender, German Radio, and the Public Sphere, 1923–1945* (Michigan: University of Michigan Press)

Lacey, K. (2013) *Listening Publics: The Politics of Listening in the Media Age* (Cambridge: Polity)

Lambert, R. (1940) *Ariel and All His Quality* (London: Gollanz)

Lang, E.M. (1929) *British Women in the Twentieth Century* (London: T. Werner Laurie)

Law, C. (1997) *Suffrage and Power: The Women's Movement 1918–1928* (London: I.B. Tauris)

Leishman, M. (2006) *My Father: Reith of the BBC* (Edinburgh: Saint Andrew Press)

LeMahieu, D.L. (1988) *A Culture for Democracy: Mass Communications and the Cultivated Mind in Britain between the Wars* (Oxford: Oxford University Press)

Lewis, C. (1924) *Broadcasting from Within* (London: George Newnes)

Lewis, J. (1984) *Women in England 1870–1950: Sexual Divisions and Social Change* (Brighton: Wheatsheaf Books)

Lewis, J. (1990) *The Politics of Motherhood: Child and Maternal Welfare in England 1900–1939* (London: Croom Helm)

Light, A. (1991) *Forever England, Femininity, Literature and Conservatism between the Wars* (London: Routledge)

Light, A. (2007) *Mrs Woolf and the Servants: The Hidden Heart of Domestic Service* (London: Fig Tree)

MacKenzie, N., and MacKenzie, J., eds. (2000) *The Diaries of Beatrice Webb* (London: Virago)

Mannin, E. (1932) *All Experience* (London: Jarrolds)

Maschwitz, E. (1957) *No Chip on My Shoulder* (London: Herbert Jenkins)

Martindale, H. (1938) *Women Servants of the State, 1870–1938* (London: Allen and Unwin)

Martindale, H. (1944) *From One Generation to Another 1839–1944: A Book of Memoirs* (London: George Allen and Unwin)

Matheson, H. (1933) Broadcasting (London: Thornton Butterworth)

McIntyre, I. (1993) *The Expense of Glory: A Life of John Reith* (London: Harper Collins)

McKibbin, R. (1998) *Classes and Cultures: England 1918–1951* (Oxford: Oxford University Press)

McNally, F. (1979) *Women for Hire: A Study of the Female Office Worker* (London: Macmillan)

Meynell, A.(1984) *Public Servant, Private Women: An Autobiography* (London: Victor Gollanz)

Miall, L. (1994) *Inside the BBC: British Broadcasting Characters* (London: Weidenfeld & Nicolson)

Mills, A.J., and Tancred, P., eds. (1992) *Gendering Organizational Analysis* (Newbury Park: Sage)

Nicholas, S. (1996) *The Echo of War: Home Front Propaganda and the Wartime BBC, 1939–1945* (Manchester: Manchester University Press)

Nicholson, V. (2007) *Singled Out: How Two Million Women Survived without Men after the First World War* (London: Viking)

Nicolson, N. (2004) *Harold Nicolson: Diaries and Letters, 1907–1964* (London: Weidenfeld and Nicolson)

Oldfield, S., ed. (1994) *This Working Day World: Women's Lives and Cultures in Britain* (London: Taylor & Francis)

Oram, A. (1996) *Women Teachers and Feminist Politics 1900–1939* (Manchester: Manchester University Press)

Oram, A. (2007) *"Her Husband Was a Woman!" Women's Gender-Crossing in Modern British Popular Culture* (London: Routledge)

Pawley, P. (1972) *BBC Engineering 1922–1972* (London: BBC Books)

Pegg, M. (1983) *Broadcasting and Society 1918–1939* (London: Croom Helm)

Reith, J. (1924) *Broadcast over Britain* (London: Hodder and Stoughton)

Reith, J. (1949) *Into the Wind* (London: Hodder and Stoughton)

Richards, J. (1984) *The Age of the Dream Palace: Cinema and Society 1930–1939* (London: Routledge)

Ross, G. (1961) *Television Jubilee, The Story of 25 Years of BBC Television* (London: W.H. Allen)

Routh, G. (1965) *Occupation and Pay in Great Britain, 1906–1960* (Cambridge: Cambridge University Press)

Rowntree, B.S. (1941) *Poverty and Progress: A Second Social Survey of York* (London: Longmans)

Russell, A. (1984) *A Certain Voice* (Bolton: Ross Anderson)

Ryan, D.S. (1997) *The Ideal Home through the 20th Century* (London, Hazar)

Scannell, P., and Cardiff, D. (1991) *A Social History of British Broadcasting, 1922–1939* (London: Basil Blackwood)

Seaton, J. (2015) *Pinkoes and Traitors: The BBC and the Nation 1974–1987* (London: Profile)

Shapley, O. (1996) *Broadcasting: A Life* (London: Scarlet Press)

Silvey, R. (1974) *Who's Listening? The Story of BBC Audience Research* (London: Allen and Unwin)

Smith, H.L. (1930–1935) *New Survey of London Life and Labour*, 9 volumes (London)

Strachey, R. (1935) *Careers and Openings for Women: A Survey of Women's Employment and a Guide for Those Seeking Work* (London: Faber and Faber)

Strachey, R., ed. (1936) *Our Freedom and Its Results by Five Women* (London: Hogarth Press)

Street, S. (2006) *Crossing the Ether: British Public Radio and Commercial Competition 1922–1945* (Eastleigh: John Libby)

Todd, S. (2005) *Young Women, Work and the Family in England 1918–1950* (Oxford: Oxford University Press)

Vicinus, M. (1985) *Independent Women: Work and Community for Single Women* (London: Virago)

Walby, S. (1986) *Patriarchy at Work: Patriarchal and Capitalist Relations in Employment* (Cambridge: Polity Press)

Wellesley, J. (2008) *Wellington: A Journey through My Family* (London: Weidenfeld & Nicolson)

Wightman, C. (1999) *More Than Munitions: Women, Work and the Engineering Industries 1900–1950* (London: Longman)

Woolf, V. (1938/2000) *Three Guineas* (Oxford: Oxford University Press)

BOOK CHAPTERS, JOURNAL ARTICLES, THESIS, UNPUBLISHED MEMOIRS

Bhatt, J. (1995) 'Margaret Miller and the Campaign for the Right of the Married Woman to Earn' (Unpublished M.Phil dissertation: University of Leicester)

Bourne, J. (2014) 'Helena Normanton and the Opening of the Bar to Women' (Unpublished doctoral dissertation: University of London)

Davy, T. 'Shorthand Typists in London 1900-1939' in Davidoff, L., and Westover, B., eds. (1986) *Our Work, Our Lives, Our Words: Women's History and Women's Work* (London: Macmillan Education) pp.124-44

Faraday, J. (2009) 'A Kind of Superior Hobby: Women Managers in the John Lewis Partnership 1918-1950' (Unpublished M.Phil dissertation: University of Wolverhampton)

Feldman, S. 'Twin Peaks: The Staying Power of BBC Radio 4's Woman's Hour' in Mitchell, C., ed. (2000) *Women in Radio: airing differences* (London: Routledge) pp.64-72

Glew, H. (2009) 'Women's Employment in the General Post Office, 1914-1939' (Unpublished doctoral dissertation: University of London)

Hackney, F. (2011) '"They Opened Up a Whole New World': Feminism, Modernity and the Feminine Imagination in Women's Magazines, 1919-1939' (Unpublished doctoral dissertation: University of London)

Hendy, D. (2012) 'Biography and the emotions as a missing 'narrative' in media history: a case study of Lance Sieveking and the early BBC', *Media History*, 18 (3–4), 361–78

Hendy, D. (2014) 'The Great War and British broadcasting: emotional life in the creation of the BBC', *New Formations*, 82, 82–99

Hunter, F. (1994) 'Hilda Matheson and the BBC 1926-1940' in Sybil Oldfield ed. *This Working Day World: Women's Lives and Cultures in Britain* (London: Taylor & Francis) pp.169-74

Jones, A. (2012) 'Mary Adams and the producer's role in early BBC science broadcasts', *Public Understanding of Science*, 21, 968–83

Michaelsen, K. (2005) 'Union Is Strength': The Medical Women's Federation and the Politics of Professionalism, 1917-30' in Cowman, K., and Jackson, L., eds. (2005) *Women and Work Culture in Britain c.1850-1950* (Aldershot: Ashgate) pp.161-176

Moores, S. (1988) '"The Box on the Dresser": Memories of Early Radio and Everyday Life', *Media, Culture and Society* 10(1) 23–40

Murphy, K. (2014) 'A Marriage Bar of Convenience? The BBC and Married Women's Work 1923-1939', *Twentieth Century British History*, Vol. 25, 4, 533–561.

Murphy, K. (2014) 'From Women's Hour to Other Women's Lives: BBC Talks for Women and the Women who Made Them, 1923-1939' in Andrews, M., and McNamara, S., eds. *Women and the Media. Feminism and Femininity in Britain, 1900 to the Present* (London: Routledge)

Sanderson, K. 'Women Civil Service Clerks 1925-1939' in Davidoff, L., and Westover, B., eds. (1986) *Our Work, Our Lives, Our Words: Women's History and Women's Work* (London: Macmillan Education) pp.145-60

Somerville, S. 'How School Broadcasting Began' in Palmer, R., ed. (1947) *School Broadcasting in Britain* (London: BBC)

Takayanagi, M. (2012) 'Parliament and Women c.1900–1945' (Unpublished doctoral dissertation: University of London)

Taylor, J. (2013) 'From Sound to Print in Pre-War Britain: the Cultural and Commercial Interdependence between Broadcasters and Broadcasting Magazines in the 1930s' (Unpublished doctoral dissertation: University of Bournemouth)

Wade, R. (no date) *Early Life at the BBC* (Unpublished memoir)

INDEX

© The Editor(s) (if applicable) and The Author(s) 2016
K. Murphy, *Behind the Wireless*,
DOI 10.1057/978-1-137-49173-2